IUTAM SYMPOSIUM ON NUMERICAL SIMULATION OF NON-ISOTHERMAL FLOW OF VISCOELASTIC LIQUIDS

FLUID MECHANICS AND ITS APPLICATIONS
Volume 28

Series Editor: R. MOREAU

MADYLAM
Ecole Nationale Supérieure d'Hydraulique de Grenoble
Boîte Postale 95
38402 Saint Martin d'Hères Cedex, France

Aims and Scope of the Series

The purpose of this series is to focus on subjects in which fluid mechanics plays a fundamental role.

As well as the more traditional applications of aeronautics, hydraulics, heat and mass transfer etc., books will be published dealing with topics which are currently in a state of rapid development, such as turbulence, suspensions and multiphase fluids, super and hypersonic flows and numerical modelling techniques.

It is a widely held view that it is the interdisciplinary subjects that will receive intense scientific attention, bringing them to the forefront of technological advancement. Fluids have the ability to transport matter and its properties as well as transmit force, therefore fluid mechanics is a subject that is particulary open to cross fertilisation with other sciences and disciplines of engineering. The subject of fluid mechanics will be highly relevant in domains such as chemical, metallurgical, biological and ecological engineering. This series is particularly open to such new multidisciplinary domains.

The median level of presentation is the first year graduate student. Some texts are monographs defining the current state of a field; others are accessible to final year undergraduates; but essentially the emphasis is on readability and clarity.

For a list of related mechanics titles, see final pages.

IUTAM Symposium on
Numerical Simulation of Non-Isothermal Flow of Viscoelastic Liquids

Proceedings of an IUTAM Symposium
held in Kerkrade, The Netherlands,
1–3 November 1993

Edited by

J. F. DIJKSMAN
Philips Research Laboratories, Eindhoven, The Netherlands

and

G. D. C. KUIKEN
Delft University of Technology, Delft, The Netherlands

SPRINGER-SCIENCE+BUSINESS MEDIA, B.V.

A C.I.P. Catalogue record for this book is available from the Library of Congress.

ISBN 978-94-010-4086-0 ISBN 978-94-011-0191-2 (eBook)
DOI 10.1007/978-94-011-0191-2

Printed on acid-free paper

CONTENTS

Preface vii
Committees viii
Participants ix

1. **J.M. Wiest. Kinetic theories for nonisothermal flows: differential constitutive equations** . . . 1

2. **B.H.A.A. van den Brule. Micro-rheological modelling of heat conduction in polymeric liquids** . . . 11

3. **G.W.M. Peters. Thermorheological modelling of viscoelastic materials** . . . 21

4. **P. Wapperom and M.A. Hulsen. Numerical simulation of a viscoelastic fluid with anisotropic heat conduction** . . . 37

5. **D.S. Malkus, J.A. Nohel and B.J. Plohr. Oscillations in piston-driven shear flow of a non-Newtonian fluid** . . . 57

6. **V. Bazin, C. Guillopé and J.-C. Saut. Some existence and stability results for nonisothermal flows of viscoelastic fluids** . . . 73

7. **M. Fillon and J. Frêne. Numerical simulation and experimental results on thermo-elasto-hydrodynamic tilting-pad journal bearings** . . . 85

8. **S. Gunter and T.N. Phillips. Non-isothermal effects in capillary viscometry** . . . 101

9. **D. Ding, P. Townsend and M.F. Webster. Numerical simulation of filling problems related to injection moulding** . . . 133

10. **X.K. Li and A.R. Davies. Numerical modelling of nonisothermal viscoelastic flow between eccentrically rotating cylinders** 159

11. **F.P.T. Baaijens. Numerical analysis of the non-isothermal viscoelastic flow of a polymer melt around a cylinder** . . . 179

Index . . . 193

PREFACE

During the last decades a considerable effort has been made on the computation of the isothermal flow of viscoelastic fluids. In fact the activities related to this particular field of non-Newtonian Fluid Mechanics have focussed on the following questions: which type of constitutive equation describes non-Newtonian fluid behaviour, how to measure fluid parameters and what type of computational scheme leads to reliable, stable and cost-effective computer codes. During the same period typical non-Newtonian fluid phenomena have been experimentally examined, such as the flow through a "four-to-one" contraction, the flow around a sphere or separation flow (see D.V. Boger and K. Walters, Rheological Phenomena in Focus, Elsevier Science Publishers 1993), providing fresh challenges for numerical modellers (consult Journal of Non-Newtonian Fluid Mechanics, Issued by Elsevier Science Publishers). Apart from momentum transport, however, fluid flow is strongly influenced by heat transport in most real industrial operations in which non-Newtonian fluids are processed.

The IUTAM-Symposium on "Numerical Simulation of Nonisothermal flow of Viscoelastic Liquids" held at ROLDUC ABBEY in Kerkrade, The Netherlands, November 1–3, 1993 has been organised to monitor the state of affairs with regard to the influence of nonisothermal effects on the flow of viscoelastic liquids.

A number of questions already dealt with during the computation of isothermal flows are still relevant and perhaps even more intriguing present when considering the computational aspects of the nonisothermal flow of viscoelastic liquids, such as:
- what type of constitutive equations describe the momentum and heat transport in a flowing viscoelastic medium,
- how to define the equation of energy, how to deal e.g. with the storage of elastic energy when the flowing viscoelastic liquid experiences a sudden change in deformation or thermal loading,
- mathematical and computational backgrounds of computer codes for calculation the nonisothermal flow of viscoelastic liquids.

It goes without saying that in order to validate the computational results well-described experiments are needed that clearly exhibit phenomena that can be attributed to thermal effects.

The present collection of papers gives an overview of what has been achieved so far. It may serve as a first milestone in a rapidly emerging and exciting new field in non-Newtonian Fluid Mechanics.

J.F. Dijksman, Eindhoven,
G.D.C. Kuiken, Delft,
The Netherlands, August 1994.

Scientific committee:

Prof. R.C. Armstrong (USA)
Prof. F.P.T. Baaijens (The Netherlands)
Dr. B.H.A.A. van den Brule (The Netherlands)
Prof. M.J. Crochet (Belgium)
Dr. J.F. Dijksman (Chairman, The Netherlands)
Dr. G.D.C. Kuiken (The Netherlands)
Prof. J.A. Nohel (USA)
Dr. J.R.A. Pearson (United Kingdom)
Prof. J.M. Piau (France)
Prof. R.I. Tanner (Australia)
Prof. K. Walters (United Kingdom)
Prof. L. van Wijngaarden (Bureau IUTAM, The Netherlands)

Local organising committee:

F.P.T. Baaijens
B.H.A.A. van den Brule
J.F. Dijksman
J.M. van der Kamp
G.D.C. Kuiken

The symposium has been sponsored by International Union of Theoretical and Applied Mechanics, by The Royal Institution of Engineers in the Netherlands and by Kluwer Academic Publishers.

List of participants:

R. Akkerman, Twente University, The Netherlands.
F.P.T. Baaijens, Philips Research Laboratories, The Netherlands.
B.H.A.A. van den Brule, Shell Research, The Netherlands.
P.C. Dauby, University of Liège, Belgium.
A.R. Davies, University of Wales, United Kingdom.
J.F. Dijksman, Philips Research Laboratories, The Netherlands.
J.B.M. Frêne, University of Poitiers, France.
M.I. Gerritsma, University of Groningen, The Netherlands.
A. Goublomme, University of Louvain-la-Neuve, Belgium.
C. Guillopé , University of Paris, France.
J. Huétink, Twente University, The Netherlands.
M.A. Hulsen, Delft University of Technology, The Netherlands.
R.J.J. Jongschaap, University of Twente, The Netherlands.
G.D.C. Kuiken, Delft University of Technology, The Netherlands.
H.E.H. Meijer, Eindhoven University of Technology, The Netherlands.
J.A. Nohel, ETH Zürich, Switzerland; University of Wisconsin, USA.
G.W.M. Peters, Eindhoven University of Technology, The Netherlands.
J. Pfragner, University of Graz, Austria.
T.N. Phillips, University of Wales, United Kingdom.
G. Rekers, Twente University, The Netherlands.
Y. Renardy, Virginia Polytechnic Institute, USA.
J.J.M. Slot, DSM-Research, The Netherlands.
R.I. Tanner, University of Sydney, Australia.
P. Wapperom, Delft University of Technology, The Netherlands.
M.F. Webster, University of Wales, United Kingdom.
J.M. Wiest, Purdue University, USA.
L. van Wijngaarden, Twente University, The Netherlands.

KINETIC THEORIES FOR NONISOTHERMAL FLOWS: DIFFERENTIAL CONSTITUTIVE EQUATIONS

JOHN M. WIEST
School of Chemical Engineering
Purdue University
W. Lafayette, IN 47907-1283 U. S. A.

Abstract. Elastic dumbbell kinetic theory has been used to derive a differential constitutive equation for the nonisothermal flow of polymer melts. This theory uses anisotropic drag to incorporate interactions between polymer molecules; the resulting constitutive equation is the nonisothermal form of the Giesekus equation. The constitutive equation thus obtained differs from that which would be obtained by applying the pseudo-time postulate to the isothermal form of the Giesekus constitutive equation. To examine this difference, we consider predictions for the inception of uniaxial elongation with simultaneous cooling at a constant rate and predictions for the inception of shear flow with simultaneous heating at a constant rate. The former experiment has been suggested as one that may be able to distinguish the applicability of the pseudo-time idea, but the results given here indicate that the latter may be more appropriate.

Keywords: Nonisothermal elongation; Nonisothermal shear; Polymer melts; Pseudo-time, Thermo-rheological simplicity

1. Introduction

All polymer processing operations involve changes in the temperature of the material, and these temperature changes can be very large and occur rapidly. In fact, given the large viscosities that typify polymer melts and the associated dissipation of mechanical energy, it is probable that few flows - including even rheometric flows - can be classified as being truly isothermal. However, almost all of the constitutive modeling that has been done for polymers has been for flows and materials at constant temperature. This dichotomy points us toward an area that obviously requires study.

The traditional mechanism for examining the temperature dependence of rheological properties is the principle of *time-temperature superposition* [1]. This principle states that all of the relaxation processes in the material have the same temperature dependence and can be described by a single function of the thermodynamic state of the material - the shift function. However, time-temperature superposition is applicable only for relating rheological properties at different constant temperatures. It does not incorporate any rheological response that results from spatial or temporal variations in temperature. A continuum mechanical framework called *thermo-rheological simplicity* that does allow for the inclusion of such temperature variations involves the use of a *pseudo-time* [2,3,4]. Within this framework, temperature variations are accounted

J.F. Dijksman and G.D.C. Kuiken (eds.),
IUTAM Symposium on Numerical Simulation of Non-isothermal Flow of Viscoelastic Liquids, 1-10.

for by dilating the time variable in the constitutive equation. That is, the time, t, in the constitutive equation is replaced by the pseudo-time, ξ, defined by:

$$d\xi = \frac{dt}{a_T[T(t)]} \tag{1}$$

or:

$$\xi = \int^{t} \frac{dt'}{a_T[T(t')]} \tag{2}$$

where a_T is the shift function that would be applied for time-temperature superposition. Although the pseudo-time idea is intuitively attractive, it has yet to be verified experimentally for polymeric liquids.

Kinetic theory provides an attractive option for exploring the behavior of macromolecular materials because it can yield very specific constitutive relations once a molecular model has been proposed. There have been several kinetic theory studies of the nonisothermal rheology of polymeric liquids. These include the works by Marrucci [5], Bird [6], and Gupta and Metzner [7] wherein dilute polymer solutions are modeled using bead-spring dumbbells. These works for dilute solutions have also been extended to bead-spring (*i.e.*, Rouse) chain models for the molecules [8]. Wiest and Phan-Thien [9] have extended the Curtiss-Bird theory [10] for molten polymers to describe nonisothermal flow. Each of the constitutive equations obtained from the theories listed above predicts deviations from the principle of thermo-rheological simplicity.

Recently, a differential constitutive equation for the nonisothermal flow of polymer melts has been proposed [11]. This constitutive equation also predicts deviations from thermo-rheological simplicity. Our purpose here is to explore the consequences of this deviation for a special case of the differential constitutive equation.

2. Constitutive Equation

The constitutive equation that we consider here is a special case of one proposed recently [11] for nonisothermal flow of polymer melts modeled as elastic dumbbells with anisotropic drag of the form suggested by Giesekus [13]. The general constitutive equation [11] uses finitely-extensible-nonlinear-elastic springs under the Peterlin approximation, but here we consider the Hookean limit of the springs. The constitutive equation is:

$$\tau + \lambda_H \tau_{(1)} - \frac{a}{nkT}\tau \cdot \tau - \lambda_H \frac{D \ln T}{Dt}\tau = -nkT\lambda_H \gamma_{(1)} \tag{3}$$

where we have used the notation of Bird, Curtiss, Armstrong, and Hassager [12]. Specifically, τ is the nonequilibrium stress (positive in compression), nkT is the product of the number density of dumbbells, Boltzmann's constant, and the absolute temperature, the subscript "(1)" indicates a convected derivative, $\gamma_{(1)}$ is the rate of strain tensor, λ_H is a temperature-dependent time constant, and a is temperature-independent parameter. For isothermal flow, this equation reduces to the Giesekus constitutive equation.

Applying the principle of thermo-rheological simplicity to the isothermal Giesekus constitutive equation gives:

$$\tau + a_T \lambda_0 \tau_{(1)} - \frac{a}{nkT} \tau \cdot \tau = -nkTa_T \lambda_0 \gamma_{(1)} \tag{4}$$

where λ_0 is the time constant evaluated at some reference temperature T_0. Equation (3) differs from Equation (4) only in the appearance of the fourth term on the left-hand side of the former. The purpose of this contribution is to explore the importance of this difference for two nonisothermal flows.

In both Equations (3) and (4) the temperature dependence of the time constant λ_H is included with a shift function a_T. That is:

$$a_T = \frac{\lambda_H(T)}{\lambda_0(T_0)} \tag{5}$$

where T_0 is a reference temperature and λ_0 is the time constant evaluated at that reference temperature. The theory does not give an explicit expression for the functional dependence of $a_T(T)$; here we simply assume [11] that it is given by:

$$a_T = \frac{T_0}{T} \exp\left[\beta\left(1 - \frac{T}{T_0}\right)\right] \tag{6}$$

where β is a parameter and T_0 is a reference temperature.

3. Nonisothermal Elongational Flow

Matsumoto and Bogue [14] suggested a nonisothermal elongational flow as a test of constitutive equations that purport to include temporal variations in temperature. In this flow, the fluid is caused to move so that its velocity is given by:

$$v_x = -\frac{1}{2}\dot{\varepsilon}x \; ; \quad v_y = -\frac{1}{2}\dot{\varepsilon}y \; ; \quad v_z = \dot{\varepsilon}z \tag{7}$$

with the elongation rate, $\dot{\varepsilon}(t)$ given by:

$$\dot{\varepsilon}(t) = \dot{\varepsilon}_0 H(t) \tag{8}$$

where $\dot{\varepsilon}_0$ is the ultimate elongation rate and $H(t)$ is the Heaviside unit step function. Simultaneously, the material is cooled so that its temperature decreases linearly with time. That is, the temperature is given by:

$$T = T_0\left(1 - Rt/\lambda_0\right) \tag{9}$$

where R is constant. The temperature is assumed to be spatially uniform within the material. The material function that is to be measured is defined by:

$$\overline{\eta}_T^+ = -\left[\tau_{zz}(t) - \tau_{xx}(t)\right]/\dot{\varepsilon}_0 \tag{10}$$

The subscript 'T' on $\overline{\eta}_T^+$ is to emphasize that the function is defined for nonisothermal flow. If the flow is isothermal, this material function reduces to the elongational viscosity growth function.

The predictions of Equations (3) and (4) for $\overline{\eta}_T^+$ for a variety of cooling rates are shown in Figure 1. Also included in the Figure is the isothermal prediction (zero cooling

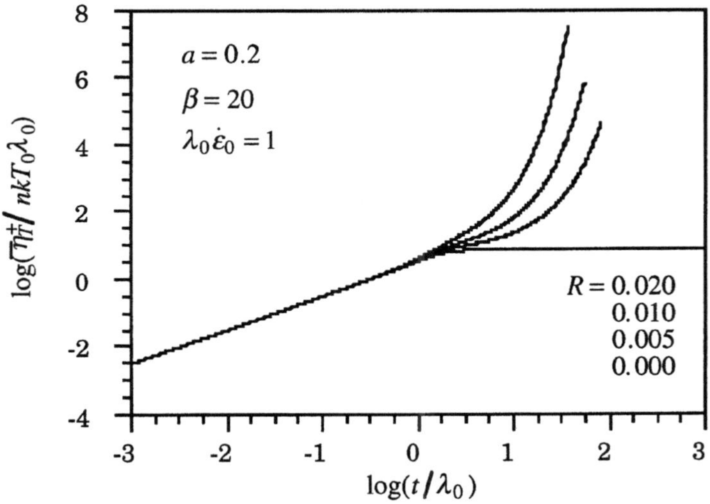

Figure 1. Predictions of the elongational viscosity during the inception of uniaxial elongational flow with simultaneous cooling at several constant rates.

rate). The predicted dependence on time and cooling rate is in at least qualitative agreement with the measurements of Matsumoto and Bogue [14]. However, the Figure illustrates the inability of this experiment to distinguish between equations (3) and (4). The predictions of both forms of the constitutive equation are very nearly identical at all but the largest cooling rate shown; the curves appear to overlap in the Figure. The relative differences between the predictions of the two forms of the constitutive equation are shown in Figure 2. Here and in what follows, we denote the predictions of the constitutive equation

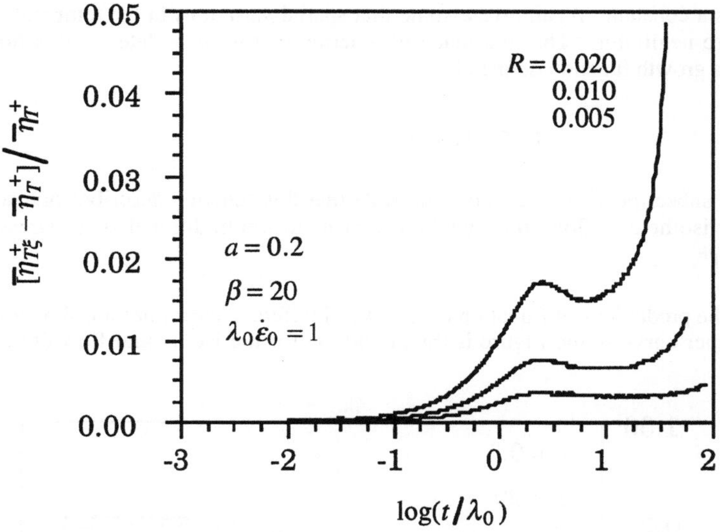

Figure 2. Relative difference in the predictions by the two variants of the constitutive equation for the inception of steady uniaxial elongational flow with simultaneous cooling at constant rates.

obtained by straightforward application of the pseudo-time postulate (Equation (4)) by the subscript ξ. The differences between the predictions of the two forms of the constitutive equation are larger at longer times and at higher cooling rates, but they are quite small.

4. Nonisothermal Shear Flow

The second nonisothermal shear flow that we consider is the inception of steady shear flow with heating at a constant rate. That is, we cause the fluid to move in such a way that its velocity is given by:

$$v_x = \dot{\gamma}(t) y \;; \quad v_y = v_z = 0 \tag{11}$$

where:

$$\dot{\gamma}(t) = \dot{\gamma}_0 H(t) \tag{12}$$

Here $\dot{\gamma}_0$ is a constant. Simultaneously, the temperature of the material is caused to increase linearly with time so that:

$$T = T_0\left(1 + \alpha t/\lambda_0\right) \tag{13}$$

where α is a constant. Again, we assume that spatial variations in the temperature in the material are negligible. The first material function that we calculate for this flow is the shear stress growth function defined by:

$$\eta_T^+ = -\tau_{xy}(t)/\dot{\gamma}_0 \tag{14}$$

Where the subscript 'T' is intended to indicate that this function describes nonisothermal flow. For isothermal flow, this function reduces to the traditional shear stress growth function, η^+.

The predictions of Equation (3) at several different shear rates are shown in Figure 3. The upper curve in the Figure is the prediction for the isothermal flow (zero heating

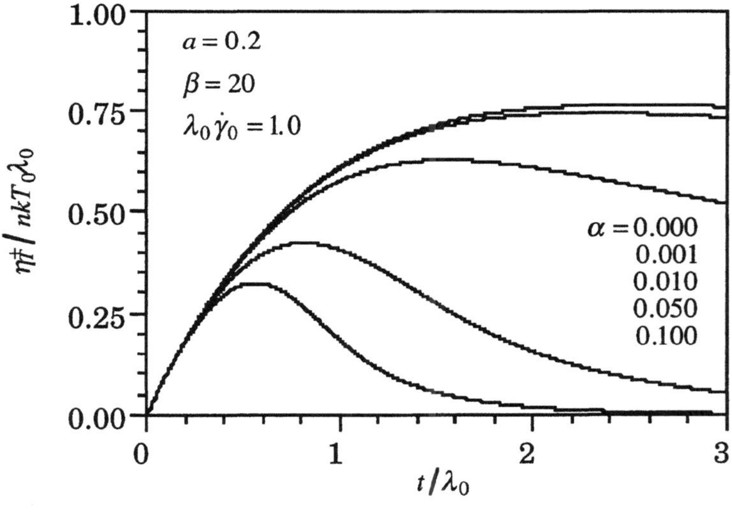

Figure 3. .Predictions of the growth of the shear viscosity during the inception of shear flow with simultaneous heating at a constant rate. The upper curve is the isothermal result.

rate). The isothermal Giesekus equation predicts a shear stress overshoot upon the inception of shear flow, but the value of the shear rate for which Figure 3 is drawn is sufficiently low that the overshoot is not easily visible in the Figure. However, at non-zero cooling rates, there is an apparent overshoot. This indicates the decrease in the shear stress caused by the continuing increase in the temperature after the initial stress growth [11].

The relative differences between the predictions of Equations (3) and (4) for the viscosity growth function with simultaneous heating are shown in Figure 4. Once again,

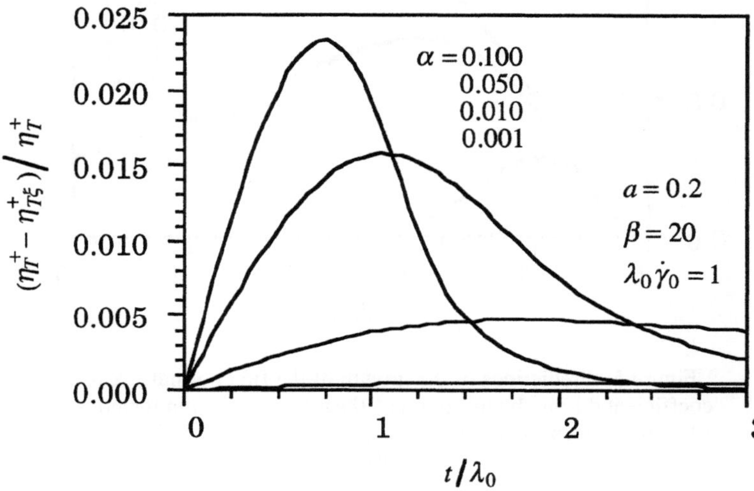

Figure 4. Relative difference in the predictions of the shear stress growth function by the two variants of the constitutive equation for the inception of steady shear flow with simultaneous heating at a variety of constant rates.

the differences are quite small. However, they are larger at shorter times and for larger heating rates.

The second material function that we calculate for the inception of shear flow with simultaneous heating at a constant rate shows the growth of the first normal stress coefficient. It is defined by:

$$\Psi_{1T}^{+} = -\left[\tau_{xx}(t) - \tau_{yy}(t)\right] / \dot{\gamma}_0^2 \tag{15}$$

Once, again, the subscript *'T"* indicates that the function is defined for the nonisothermal flow. Under isothermal conditions, the function reduces to the first normal stress growth function, Ψ_1^+. The predictions of Equation (3) for this function are shown in Figure 5 for a variety of heating rates. The isothermal Giesekus equation does predict an overshoot in the first normal stress difference, but the overshoot is quite small at the shear rate for which Figure 5 is drawn. At non-zero heating rates, there is an apparent overshoot in the normal stress coefficient. The relative differences between the predictions of Equations (3)

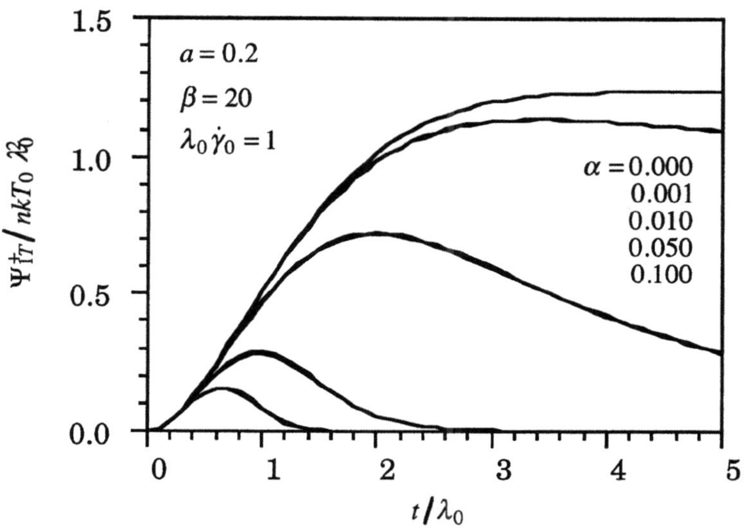

Figure 5. Predictions of the growth of the first normal stress coefficient during the inception of shear flow with simultaneous heating at a constant rate. The upper curve is the isothermal result.

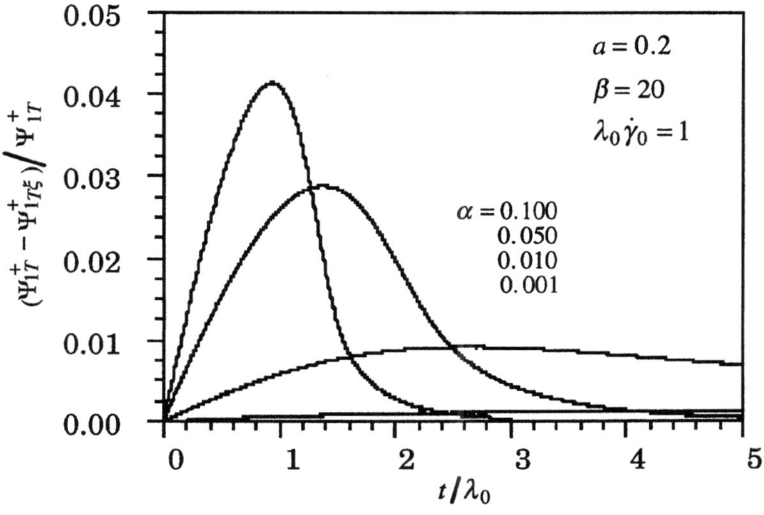

Figure 6. Relative difference in the first normal stress growth function for the two forms of the constitutive equation

and (4) is shown in Figure 6. The differences in the first normal stress growth function are qualitatively similar to those for the shear stress growth function, and they show the same trends. However, the magnitudes of the former are about twice those of the latter.

5. Conclusions

We have considered a constitutive equation for nonisothermal flow derived from dumbbell kinetic theory and the same constitutive equation for isothermal flow modified so as to describe nonisothermal flow by way of the pseudo-time postulate. These two constitutive equations are different, thereby casting doubt upon the validity of the pseudo-time postulate. The predictions of these constitutive equations for two types of nonisothermal flow have also been presented. The first flow that we considered is the inception of uniaxial elongation with cooling at a constant rate as suggested by Matsumoto and Bogue [14]. Our results indicate that the predictions of the two nonisothermal variants of the constitutive equation are qualitatively similar to the data of Matsumoto and Bogue [14]. Furthermore, the predictions of the two forms of the constitutive equation are quite similar for this flow differing by a maximum of about five percent for the values of the parameters considered. The second type of flow that we considered is the inception of steady shear flow with heating at a constant rate. Unfortunately, data for such a flow are not available, but it may be feasible to design an appropriate experiment to obtain them. Both forms of the constitutive equation gave very similar predictions for the growth of both the shear stress and the first normal stress difference in this flow. The relative variations are greater for the first normal stress difference than for the shear stress, but they are still only about five percent for the values of the parameters that were considered.

For both of the flows considered, the relative differences between the predictions of the two constitutive equations become smaller at long times. This is not as severe a limitation as it might at first seem. Any experiment that is to test the pseudo-time idea must include temperature transients, and, therefore, the experiment cannot reach a steady state. Consequently, experiments to test nonisothermal constitutive equations must be designed to distinguish differences in temporal responses.

The results obtained here are encouraging in that they indicate that, for the flows considered, the differences between the constitutive equation obtained from dumbbell kinetic theory and that obtained by application of the pseudo-time postulate do not lead to substantial differences in the predictions of material functions. However, this should not lead one to conclude that the predictions of the constitutive equations for more complicated nonisothermal flows will be similar. The same results are discouraging in that they indicate that it will not be straightforward to test between the two forms of the nonisothermal constitutive equation using the inception of steady shear or the inception of steady elongation. Nonetheless, given that the relative material function differences for the two constitutive equations in shear and elongational flows are of similar magnitude, and that the shear flow experiment should be easier to perform, we are lead to conclude that shear flow with simultaneous heating may be a good candidate for much-needed nonisothermal experimentation.

6. Acknowledgment

The author gratefully acknowledges financial support provided by the 3M Company. In addition, he would like to thank Dr. G. D. C. Kuiken for pointing out an error in an early version of the manuscript.

7. References

1. Ferry, J. D.: 1980 *Viscoelastic Properties of Polymers* , 3rd Edition, John Wiley and Sons, New York
2. Morland, L. W. and Lee, E. H.: 1960, 'Stress Analysis for Linear Viscoelastic Materials with Temperature Variation', *Trans. Soc. Rheol.*, **4**, 233-263
3. Crochet, M. J., Naghdi, P.M.: 1969, 'A Class of Simple Solids with Fading Memory', *Int. J. Eng. Sci.*, **7**, 1173-1198
4. Tanner, R. I.: 1985, *Engineering Rheology*, Oxford Press, Oxford
5. Marrucci, G: 1972, 'The Free Energy Constitutive Equation for Polymer Solutions from the Dumbbell Model', *Trans. Soc. Rheol.*, **16**, 321-330
6. Bird, R. B.:1979, 'Use of Simple Molecular Models in the Study of Mechanical Behavior of Solutions of flexible Macromolecules', *J. Non-Newtonian Fluid Mech.*, **5**, 1-12
7. Gupta, R. K. and Metzner, A. B., 1982, 'Modeling of Nonisothermal Polymer Processes', *J. Rheol.*, **26**, 181-198
8. Wiest, J. M., 1988, 'Time-Temperature Superposition in Non-Isothermal Flow', *J. Non-Newtonian Fluid Mech.*, **27**, 127-131
9. Wiest, J. M. and Phan-Thien, N.: 1988, 'Nonisothermal Flow of Polymer Melts', *J. Non-Newtonian Fluid Mech.*, **27**, 333-347
10. Curtiss, C. F. and Bird, R. B.: 1981 'A Kinetic Theory for Polymer Melts. I. The Equation for the Single Link Orientational Distribution Function', *J. Chem. Phys.*, **74**, 2016-2025 ; 'A Kinetic Theory for Polymer Melts. II. The Stress Tensor and the Rheological Equation of State', *J. Chem. Phys.*, **74**, 2026 - 2033
11. Wiest, J. M. (1993), 'Nonisothermal Flow of Polymer Melts: A Differential Constitutive Equation' *Rheol. Acta*, submitted
12. Bird, R. B., Curtiss, C. F., Armstrong, R. C., and Hassager, O.,:1987 *Dynamics of Polymeric Liquids, Volume 2, Kinetic Theory*, 2nd Edition, Wiley -Interscience, New York
13. Giesekus, H.: 1966, 'Die Elastizität von Flüssigkeiten', *Rheol. Acta* , **5**, 29-36
14. Matsumoto T., Bogue T. C.: 1977, 'Nonisothermal Rheological Response During Elongational Flow', *Trans. Soc. Rheol.*, **21**, 453-468

MICRO-RHEOLOGICAL MODELLING OF HEAT CONDUCTION IN POLYMERIC LIQUIDS

B.H.A.A. VAN DEN BRULE
Shell Research B.V., P.O. Box 60, 2280AB-Rijswijk, The Netherlands

Abstract. In a flowing polymeric liquid molecular orientation will give rise to anisotropic conduction of heat. In this paper two models to relate the thermal conductivity tensor to the deformation history of an amorphous polymeric material are presented. The first model is based upon the network theory for polymers and the second model is based upon kinetic theory. Despite the fact that the models are relatively simple there is good agreement with the sparse experimental results.

Key words: anisotropic heat conduction, network theory, kinetic theory

1. Introduction

In high speed polymer processing and in lubrication systems, because of the high viscosity of the fluid and/or the large velocity gradients, the temperature rise due to the viscous dissipation of energy may be significant. This is especially the case for polymer melts since these fluids are very poor heat conductors which leads to local accumulation of heat. Moreover, there is the problem that in general the viscosity of a fluid is very sensitive to changes of the temperature. Therefore, not only will we have a non-uniform temperature distribution in the fluid, but this will in turn also change the flow field. For injection moulding in particular it has been shown that the result of a numerical simulation of the process is highly dependent on the value of the thermal conductivity. To date, relatively little attention has been paid to the thermal properties of polymeric liquids. Most of the available thermal conductivity data have been obtained on non-oriented samples and even in this case it is very difficult to attain an accuracy better than 5%. There is, however, evidence that molecular orientation may lead to an increase of the conductivity in the direction of the orientation with a corresponding decrease in the direction perpendicular to the orientation [1, 2, 3]. Since the reported deviations of the thermal conductivity from the equilibrium state are large enough to significantly influence the flow of a melt, it is of importance to construct a theory to relate the thermal conductivity tensor to the deformation history of a polymeric material. In this contribution two ways to analyse the thermal conductivity of a polymeric material will be presented.

A well known way of modelling rubbers and polymer melts is the concept of the polymer network. In this concept the material is considered as a three dimensional elastic network of macromolecules which are connected to each other at junctions. In a rubber the junctions are the chemical cross-links between the molecules and in a polymer melt the junctions are of temporary nature since they are formed by entanglements between the long molecules. Using this idea it is possible to construct a theory for the heat conduction where we use the fact that energy is transported

J.F. Dijksman and G.D.C. Kuiken (eds.),
IUTAM Symposium on Numerical Simulation of Non-isothermal Flow of Viscoelastic Liquids, 11-20.

more readily along the backbones of the molecules than in the direction across, from one molecule to another. The reason is that the interaction between two molecules are governed by the relatively weak van der Waals interactions, whereas the interaction between two adjacent segments of the same molecule are governed by the strong chemical bonds.

Another, for polymeric liquids perhaps more fundamental way, to model the heat conduction is to consider the thermal motion of a bead-spring chain (which mimics a macromolecule) in a solvent. In the presence of a temperature gradient in the solvent different beads of the chain will attain a different velocity distribution. Since the beads are connected by elastic springs there will be a net transport of energy between the beads under these circumstances. In a flowing solvent the viscous forces working on the chain will cause orientation and thus the thermal conductivity tensor will become anisotropic.

It is interesting to note that under certain assumptions both approaches mentioned above will result in a direct proportionality of the thermal conductivity tensor to the stress tensor.

2. Network theory for the thermal conductivity

Microscopically, heat conduction is the transfer of kinetic energy from one atom or molecule to another. The theory presented in this section is based upon the network concept for rubber and polymer melts. A comprehensive review of network theory can be found in the excellent monograph of Treloar [4] or in the textbooks written by Bird et al. [5]. In the context of network theory we distinguish between two different mechanisms that contribute to heat conduction:

- Energy transfer by means of Van der Waals interactions between neighbouring molecules. Since these interactions have a very short range we assume that this results in a non-orientable isotropic contribution to the thermal conductivity tensor, $\lambda^{vw}\mathbf{1}$.
- Energy transfer via the backbones of the molecules and by means of the strong inter-molecular interactions that serve as the junction points between different molecules. A junction may be a chemical cross-link in a rubber or an entanglement in a polymer melt. Deformation of the network will make this contribution anisotropic.

Regarding the amount of energy φ flowing through a network strand in unit time we postulate that this will be proportional to the temperature difference between the end-points of the network strand under consideration. If the length and orientation of a network segment are specified by the vector $\boldsymbol{Q}$ we may write

$$\varphi(\boldsymbol{Q}) = -\beta \nabla T \cdot \boldsymbol{Q}. \tag{1}$$

The network contribution to the thermal conductivity tensor is calculated by an analysis of the heat flux through a surface A with normal $\boldsymbol{n}$. The total flux Φ through A is found by integrating the contributions of the individual segments which intersect this plane

$$\Phi = \int \Psi(\boldsymbol{Q},t)\varphi(\boldsymbol{Q})\boldsymbol{Q}d^3\boldsymbol{Q} \cdot \boldsymbol{n}A$$

$$\Phi = -\int \Psi(\boldsymbol{Q},t)\beta\nabla T \cdot \boldsymbol{QQ}d^3\boldsymbol{Q} \cdot \boldsymbol{n}A, \tag{2}$$

where $\Psi(\boldsymbol{Q},t)d^3\boldsymbol{Q}$ is the number of segments in a unit volume with a segment vector in the range $d^3\boldsymbol{Q}$ about $\boldsymbol{Q}$. According to Fourier's law of heat conduction the heat flux through a surface A is given by:

$$\Phi = -\boldsymbol{\lambda} \cdot \nabla T \cdot \boldsymbol{n}A. \tag{3}$$

It follows from a comparison of the last two equations that the thermal conductivity tensor $\boldsymbol{\lambda}$ is given by

$$\boldsymbol{\lambda} = \lambda^{vw}\mathbf{1} + n\beta\langle\boldsymbol{QQ}\rangle, \tag{4}$$

where we also incorporated the contribution of the Van der Waals interactions and where we implicitely assumed thay β is independent of $\boldsymbol{Q}$. The angular brackets indicate an average over the configuration distribution Ψ/n and n denotes the number density of segments. At equilibrium the contribution of the network can be written as

$$\lambda^{nw} = n\beta\langle\boldsymbol{QQ}\rangle_0. \tag{5}$$

From the requirement that the equilibrium system should obey the Boltzmann distribution, the equation can be reduced further to give $\lambda^{nw} = n\beta kT/H$ if we adopt the commonly made assumption that the network segment act as linear springs with a spring constant H.

It is convenient to introduce the extra thermal conductivity tensor $\boldsymbol{\kappa}$ as a measure for the deviation from the isotropic equilibrium state in a deformed material. Accordingly we write $\boldsymbol{\lambda} = \lambda_0\mathbf{1} + \boldsymbol{\kappa}$, where

$$\lambda_0 = \lambda^{vw} + \lambda^{nw} \tag{6}$$

and

$$\boldsymbol{\kappa} = n\beta\langle\boldsymbol{QQ}\rangle - \frac{n\beta kT}{H}\mathbf{1}. \tag{7}$$

At this point it is interesting to note that the expression for the extra stress tensor for this type of network can be written as [5]:

$$\boldsymbol{\tau} = -nH\langle\boldsymbol{QQ}\rangle + nkT\mathbf{1} \tag{8}$$

A comparison of Eqs. (7) and (8) shows that the stress tensor and the thermal conductivity tensor, in analogy to the stress tensor and the refractive index tensor are linearly related:

$$\boldsymbol{\kappa} = -(\beta/H)\boldsymbol{\tau}. \tag{9}$$

If this stress-thermal relation is confirmed by experiments, it would be of great practical importance for two reasons: (i) only one extra experiment is needed in order to fully characterise the thermal conductivity tensor, and (ii) it is a very convenient expression to use in a numerical simulation.

2.1. Rubber

In order to calculate the heat conduction in a deformed material it is necessary to find the expression which relates the configuration distribution function to the deformation history of the material. For a permanently cross-linked rubber this is relatively simple since no junctions are lost nor created while deforming the material. Using the theory of rubber elasticity [4] one can show that linear segments transform affinely from the equilibrium state to the deformed state. Hence, a network segment $\boldsymbol{Q}(t)$ in the deformed state is related to the equilibrium state $\boldsymbol{Q}_0$ by

$$\boldsymbol{Q}(t) = \boldsymbol{F}(t) \cdot \boldsymbol{Q}_0, \tag{10}$$

where the tensor $\boldsymbol{F}$ is the deformation gradient. The expression for the thermal conductivity of a rubber follows readily from this transformation and Eq.(4):

$$\begin{aligned} \boldsymbol{\lambda} &= \lambda^{vw}\mathbf{1} + n\beta \boldsymbol{F} \cdot \langle \boldsymbol{QQ} \rangle_0 \cdot \boldsymbol{F}^\dagger \\ &= \lambda^{vw}\mathbf{1} + \lambda^{nw}\boldsymbol{B}, \end{aligned} \tag{11}$$

where $\boldsymbol{B}$ is the Finger strain tensor, $\boldsymbol{B} = \boldsymbol{F} \cdot \boldsymbol{F}^\dagger$. If we non-dimensionalise the thermal conductivity with its equilibrium value, we then obtain:

$$\boldsymbol{\lambda}/\lambda_0 = (1 - C)\mathbf{1} + C\boldsymbol{B}, \tag{12}$$

where the parameter C is introduced as the fraction of the heat that is conducted through the network, $C = \lambda^{nw}/\lambda_0$.

As an example we will consider uniaxial and biaxial stretching of rubber. In uniaxial elongation a rubber strip with equilibrium length L_0 is stretched to a length L in the z-direction; the strip is free to contract in the other two directions. The elongation ε is defined as $\varepsilon = L/L_0$. For this deformation the Finger strain tensor becomes:

$$\boldsymbol{B} = \begin{pmatrix} \varepsilon^{-1} & 0 & 0 \\ 0 & \varepsilon^{-1} & 0 \\ 0 & 0 & \varepsilon^2 \end{pmatrix}. \tag{13}$$

In Figure 1 the theoretical predictions for the increase of the thermal conductivity in the direction of the elongation are compared with experimental results obtained by Tautz [1]. It can be seen that good agreement is obtained.

In biaxial elongation, a sheet of rubber situated in the xy-plane, is stretched by an amount ε in both the x and y-direction while it is free to contract in the z-direction. For this deformation the Finger tensor becomes:

$$\boldsymbol{B} = \begin{pmatrix} \varepsilon^2 & 0 & 0 \\ 0 & \varepsilon^2 & 0 \\ 0 & 0 & \varepsilon^{-4} \end{pmatrix}. \tag{14}$$

In Figure 2 the decrease of the thermal conductivity in the z-direction is compared with the experiments of Hands [2]. In this case as well a good fit to the experimental results could be obtained. It should be noted however that the value of C is different in both experiments.

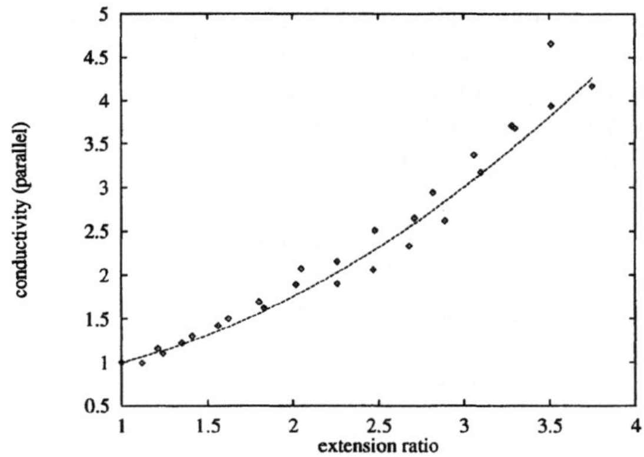

Fig. 1. The increase of the thermal conductivity in the direction of elongation in uniaxially elongated natural rubber. The experimental data are obtained by Tautz. The calculated curve is fitted to the data with a value $C = 0.25$

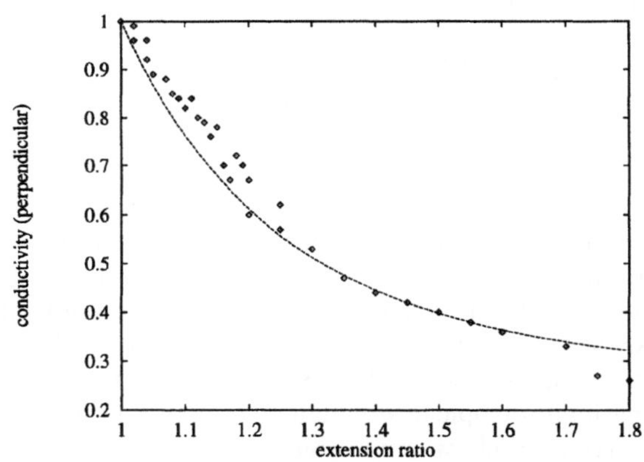

Fig. 2. The decrease of the thermal conductivity in the direction perpendicular to the plane of a biaxially extended sheet of natural rubber. The experimental data are obtained by Hands. The calculated curve is fitted to the data with a value $C = 0.75$

Unfortunately, there are only few experiments where both the increase of the heat conduction in the direction of elongation and the corresponding decrease in the direction across, are measured on the the same sample of material. One example is the experiment by Hellwege et al. [3] who measured the anisotropy of the heat conduction in polymer samples which were elongated at a temperature just above the glass transition. Subsequently, they quenched the sample to room temperature and measured the heat conduction in different directions. Also these experiments can be described accurately with the expressions given above [6].

2.2. Polymer melt

In a polymer melt the network is constantly being renewed since network segments are created and destructed. If we assume that the rate of creation of a segments, g, and the probablity of a segment being destroyed, h, are independent of the configuration $\boldsymbol{Q}$ of the segment under consideration, then the following expression for the network contribution to the thermal conductivity tensor may be derived [6, 7]:

$$\boldsymbol{\lambda}^{nw} = \int_{-\infty}^{t} \mu(t,t')\boldsymbol{B}(t,t')dt', \tag{15}$$

where the memory function $\mu(t,t')$ is given by

$$\mu(t,t') = \lambda^{nw} h e^{-(t-t')h}, \tag{16}$$

and λ^{nw} is given by

$$\lambda^{nw} = \frac{g\beta kT}{hH}. \tag{17}$$

The memory function expresses the fact that recent deformations are more important than old deformations and it gives rise to relaxation pheneomena.

In steady shear flow, where the velocity field is given by $v_x = \dot{\gamma}y, v_y = v_z = 0$, the non-zero components of the thermal conductivity tensor are given by:

$$\begin{aligned}
\lambda_{xx}/\lambda_0 &= 1 + 2\dot{\gamma}^2/h^2 \\
\lambda_{xy}/\lambda_0 &= \dot{\gamma}/h \\
\lambda_{yy}/\lambda_0 &= 1 && (18) \\
\lambda_{zz}/\lambda_0 &= 1 && (19)
\end{aligned}$$

In this example we might, for instance, consider a temperature gradient in the y-direction. In that case, owing to the fact that $\lambda_{xy} \neq 0$, the heat flux vector will no longer coincide with the direction of the temperature gradient. The increase of λ_{xx} with shear rate may have a significant effect on the length of a thermal entrance region as is demonstrated in Ref. [8] for the Graetz-Nusselt problem. Other flow calculations with anisotropic heat conduction are performed by Wapperom and Hulsen [9] and are reported elsewhere in this proceedings.

3. Non-isothermal dumbbell

In this section an extension of existing kinetic theories to non-isothermal conditions will be presented. An extensive review of kinetic theories can be found in Ref. [5]. In most kinetic theories a macromolecule is modelled as a bead-rod or bead-spring chain. In existing theories, for isothermal situations, it is usually assumed that the velocity distribution of the beads is Maxwellian about the velocity of the center-of-mass of the chain. In the presence of a temperature gradient the velocity distribution of the beads will change and will be different for different beads. It will be shown that this results into a net transfer of energy along the backbone of the chain. In case the molecules are oriented in a flowing liquid this will give rise to an anisotropic conduction of heat.

The simplest model for a molecule is a Hookean dumbbell which consists of two Brownian beads connected by a linear spring. The beads have a mass m and a friction coefficient with the environment which is equal to ζ. The spring constant is denoted by H. It is assumed that there is no appreciable change in the position of the beads on the time scale required to equilibrate the velocity distribution of the beads. In this way we may regard the environment of a bead (local solvent temperature and velocity) to be stationary when we calculate the velocity distribution of the beads. Following the theory of Brownian motion as presented by Chandrasekhar [10] one can write down the Langevin-equations for the beads:

$$m d^2 \boldsymbol{r}_i/dt^2 = -\zeta(d\boldsymbol{r}_i/dt - \boldsymbol{v}_i) \pm H\boldsymbol{Q} + \boldsymbol{f}_i^{(s)}(t), \qquad (i=1,2), \tag{20}$$

where $\boldsymbol{r}_i$ is the position of a bead, $\boldsymbol{Q} = \boldsymbol{r}_2 - \boldsymbol{r}_1$ and $\boldsymbol{f}_i^{(s)}(t)$ is a stochastic force that represents the thermal agitation of the beads by the surrounding solvent molecules. Regarding the stochastic force, we make the usual assumptions, i.e. it has a zero mean and is uncorrelated at the timescale of the motion of the beads [11]:

$$\begin{aligned} \langle \boldsymbol{f}_i^{(s)}(t) \rangle &= \boldsymbol{0}, \\ \langle \boldsymbol{f}_i^{(s)}(t+\tau)\boldsymbol{f}_i^{(s)}(t) \rangle &= 2\zeta kT\delta(\tau)\boldsymbol{1}, \\ \langle \boldsymbol{f}_1^{(s)}(t)\boldsymbol{f}_2^{(s)}(t) \rangle &= \boldsymbol{0}. \end{aligned} \tag{21}$$

The Langevin equation can be solved to give the following expression for the variance of the velocity distribution of the beads [12]

$$\sigma_i^2 = \frac{k}{m}\left(T - \frac{\nabla T \cdot \boldsymbol{Q}}{2(1 \mp mH/\zeta^2)}\right), \qquad (i=1,2), \tag{22}$$

where T is the temperature at the center of the dumbbell. Note that the difference between the velocity distribution of the two beads depends on the dimensionless group mH/ζ^2; for strongly interacting beads, $mH/\zeta^2 \gg 1$, the beads will have the same velocity distribution, whereas for $mH/\zeta^2 \ll 1$ the beads will equilibrate at the local solvent temperature T_i. From an energy balance it follows directly that the amount of energy, φ, flowing in unit time from bead "1" to bead "2" equals the sum of the work done by the stochastic and the viscous forces in unit time on bead "1". The work done by the stochastic force is independent of the velocity distribution of the beads and is given by $W_i^s = 3\zeta kT_i/m$. The work done by the viscous forces depends on the velocity distribution and is given by $W_i^v = -3\zeta\sigma_i^2$. The energy flux between the beads is thus found to be [13]:

$$\begin{aligned} \varphi &= W_i^s + W_i^v \\ \varphi &= -\frac{3\zeta k \nabla T \cdot \boldsymbol{Q}}{2m}\frac{mH/\zeta^2}{1+mH/\zeta^2}. \end{aligned} \tag{23}$$

Once the velocity distribution and the corresponding energy flux between the beads are known,the expression for the thermal conductivity tensor is obtained in a way similar to the one presented in Section 2. The formal expression for the thermal conductivity tensor is thus found to be:

$$\boldsymbol{\lambda} = \lambda^s \mathbf{1} + \frac{3n\zeta k}{2m} \frac{mH/\zeta^2}{1+mH/\zeta^2} \langle \boldsymbol{QQ} \rangle, \tag{24}$$

where λ^s denotes the conductivity of the solvent and n is the number density of dumbbells.

Using the Boltzmann distribution to evaluate $\langle \boldsymbol{QQ} \rangle_0$, we obtain for the equilibrium value of the thermal conductivity,

$$\lambda_0 = \lambda^s + \lambda^d, \tag{25}$$

and for the extra thermal conductivity,

$$\boldsymbol{\kappa} = \frac{3n\zeta k}{2mH} \frac{mH/\zeta^2}{1+mH/\zeta^2} \langle \boldsymbol{QQ} \rangle - \lambda^d \mathbf{1}, \tag{26}$$

where λ^d is given by

$$\lambda^d = \frac{3n\zeta k^2 T}{2mH} \frac{mH/\zeta^2}{1+mH/\zeta^2}. \tag{27}$$

Comparing the expression for the extra thermal conductivity $\boldsymbol{\kappa}$ with the expression for the polymer contribution to the stress tensor [5]

$$\boldsymbol{\tau}_p = -nH\langle \boldsymbol{QQ} \rangle + nkT\mathbf{1}, \tag{28}$$

we once again arrive at a stress-thermal relation in analogy to the result obtained in the network model

$$\boldsymbol{\kappa} = -\frac{3\zeta k}{2mH} \frac{mH/\zeta^2}{1+mH/\zeta^2} \boldsymbol{\tau}_p. \tag{29}$$

Using the constitutive equation derived from the Hookean dumbbell model and the stress-thermal relation, we can directly write the relation between the thermal conductivity and the kinematics of the flow field:

$$\boldsymbol{\kappa} + \tau_H \overset{\nabla}{\boldsymbol{\kappa}} = 2\tau_H \lambda^d \boldsymbol{D}, \tag{30}$$

where $_\nabla$ denotes the upper convected time derivative, $\tau_H = \zeta/4H$ is the relaxation time and $\boldsymbol{D}$ is the rate-of-strain tensor. This expression is based on the assumption that the temperature gradient has a negligible effect on the evolution of the configuration distribution function.

It is obvious that also other rheological models can be used to calculate the thermal conductivity. As an example of a relatively succesful model we will consider the Giesekus model. Giesekus proposed a constitutive equation based upon an anisotropic drag force. The idea being that in a polymer melt, or concentrated solution, a molecule is surrounded by other oriented molecules which causes the effective drag force to be anisotropic. The Giesekus equation is:

$$\boldsymbol{\tau} + \tau_H \overset{\nabla}{\boldsymbol{\tau}} - \frac{\alpha \tau_H}{\eta} \boldsymbol{\tau}^2 = -2\eta \boldsymbol{D}, \tag{31}$$

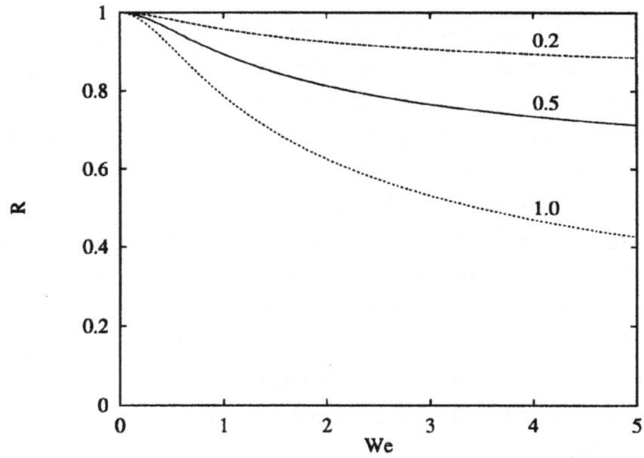

Fig. 3. The reduction of the heat conduction in the gradient direction in steady shear flow. The curve is calculated on the basis of the Giesekus model, with $\alpha = 0.5$, for $C = 1, 0.5$ and 0.2

where α is a measure for the anisotropy of the friction. When $\alpha = 0$ the friction is isotropic and the Hookean dumbbell model is recovered. The maximum value of α should not exceed $\alpha = 1$. If we assume that indeed the drag coefficient is anisotropic on the time scale of configurational changes but can be considered isotropic on the short timescale of equilibration of the velocity distribution we may use the stress-thermal relation to find the thermal conductivity.

As an example we will consider the reduction of the thermal conductivity in the gradient direction in steady shear flow. A shear flow is generated in a fluid confined between two parallel surfaces which are kept at a different temperature. By measuring the heat flux between the surfaces the reduction $R = \lambda_{yy}/\lambda_0$ of the thermal conductivity can be determined as a function of shear rate. This reduction can be written as:

$$R = \frac{\lambda_0 + \kappa_{yy}}{\lambda_0} = 1 + C\frac{\kappa_{yy}}{\lambda^d}, \tag{32}$$

where we introduced $C = \lambda^d/\lambda_0$. Using the stress-thermal relation, Eq.(29) and the Giesekus model, with $\alpha = 1/2$, this can be written as:

$$R = 1 + C\frac{\tau_{yy}}{nkT} = 1 - C + C\sqrt{2}/\sqrt{1+X}, \tag{33}$$

where

$$X = \sqrt{1 + 4We^2}. \tag{34}$$

The Weissenberg number (We) is defined by $We = \tau_H\dot{\gamma}$. In Figure 3 this result is plotted for a few values of the parameter C. Since C is probably a small number, the expected deviations of the thermal conductivity in the gradient direction will be small. Recently, Ponomarev et al. [14] performed some experiments and indeed found reductions by 5-10%.

4. Conclusion

In this paper two ways to analyse the influence of deformation on the thermal conductivity of a polymeric material are presented. It is shown that the thermal conductivity tensor can be related to the history of deformation of the material using the methods which are commonly applied in micro-rheology. For the simple models used here there appears to be a direct proportionality between the thermal conductivity tensor and the stress tensor. The existence of such a relation, in analogy to the stress-optical rule, would be of great practical importance, since only one extra experiment (to determine the proportionality constant) would be required in order to fully specify the thermal conductivity tensor. Furthermore, if such a relation exists it would be a straightforward matter to incorporate anisotropic heat conduction in numerical codes that already calculate the stress tensor. Though the models used in this paper are simple, they can be extended to more complex rheological models without much difficulty.

References

1. H. Tautz, *Bestimmung der Wärmeleitfähigkeit von Kautschukvulkanisaten in Abhängigkeit von der Dehnung*, Exper. Tech. der Phys., **7**, 1959.
2. D. Hands, *The effect of biaxial orientation on the thermal conductivity of vulcanised and unvulcanised rubber*, Rubber Chem. Technol., **53**, 1980.
3. K.-H. Hellwege, J. Hennig and W. Knappe, *Anisotropie der Wärmeausdehnung und Wärmeleiting in einachsig verstreckten amorphen Hochpolymeren*, Kolloid Z. Z. Polymere, **188**, 1963.
4. L.R.G. Treloar, *The physics of rubber elasticity*, Clarendon Press, Oxford, 1975.
5. R.B. Bird, C.F. Curtiss, R.C. Armstrong and O. Hassager, *Dynamics of polymeric liquids*, vol. 2, Kinetic theory, Wiley, New York, 1977.
6. B.H.A.A. van den Brule, *A network theory for the thermal conductivity of an amorphous polymeric material*, Rheol. Acta, **28**, 1989.
7. B.H.A.A. van den Brule and S.B.G. O'Brien, *Anisotropic conduction of heat in a flowing polymeric material*, Rheol. Acta, **29**, 1990.
8. B.H.A.A. van den Brule and P.J. Slikkerveer, *Anisotropic conduction of heat caused by molecular orientation in a flowing polymeric liquid*, Rheol. Acta, **29**, 1990.
9. P. Wapperom and M.A. Hulsen, *Numerical simulation of a viscoelastic liquid with anisotropic heta conduction*, IUTAM symposium on numerical simulation of non-isothermal flow of viscoelastic fluids, Eds. Dijksman et al., Kluwer, Dordrecht, 1993.
10. S. Chandrasekhar, *Stochastic problems in physics and astronomy*, Rev. Mod. Phys., **15**, 1943.
11. E.J. Hinch, *Application of the Langevin equation to fluid suspensions*, J. Fluid Mech., **72**, 1975.
12. B.H.A.A. van den Brule, *A contribution to the micro-rheological modelling of transport properties*, PhD thesis, Twente University, 1991.
13. B.H.A.A. van den Brule, *The non-isothermal dumbbell: a model for the thermal conductivity of a polymer solution*, Rheol. Acta, **29**, 1990.
14. S.V. Ponomarev, private communication, 1993.

THERMORHEOLOGICAL MODELLING OF VISCOELASTIC MATERIALS

GERRIT W.M PETERS
Centre for Polymers and Composites
Faculty of Mechanical Engineering,
Eindhoven University of Technology
P.O. Box 513, 5600 MB Eindhoven, The Netherlands

Abstract. A semi-empirical framework for the thermorheological description of the flow of viscoelastic materials is presented. The framework summarizes existing constitutive models and offers the possibilty to investigate and extend these models in a systematic way. Modelling starts with an expression for the stress tensor in terms of the microstructure and the principal nonaffine motion thereof. Finally, however, equations are formulated in macroscopic terms and constitutive equations of the differential type emerge. Thus it is prevented that these equations become too complex for use in numerical simulations of nonisothermal transient viscoelastic flows.

The Clausius-Duhem inequality and the energy equation are discussed in terms of this framework, allowing to decide directly whether a model is thermodynamically admissible. It is shown as an example that the differential form of the Doi/Edwards model does not fit well in the general framework of thermodynamics and suggestion to solve this problem are discussed.

1. Introduction

To describe the rheological behaviour of viscoelastic materials, a large number of constitutive equations is available [4]. Usually, these models are limited to isothermal flows and the extension to non-isothermal flows is not trivial [5]. In order to specify which part of the supplied mechanical power is dissipated and which part is accumulated as elastic energy, the energy equation needs to be reconsidered. The two limiting cases are ideal-elastic and viscous material behaviour. Deformation of elastic materials is a reversible process: mechanical energy is stored and can be released into mechanical energy again. Deformation of purely viscous materials is irreversible: mechanical energy is entirely dissipated. For viscoelastic materials, mechanical energy is partly stored in elastic energy and partly dissipated. In addition to the right partitioning of elastically stored and dissipated energy, two different ways of storing elastic energy in a viscoelastic material have to be distinguished: entropy and energy elasticity [7]. The first is related to the entropy part of the Helmholtz free energy, the second to the internal energy part.

Existing constitutive models, including those based on detailed modelling of the microstructure and therefore sometimes very complex, have no sufficient predictive capacity for complex flows of polymer melts and polymer solutions. Besides, constitutive equations that are rigorous and more accurate are in general unwieldy and impracticable for numerical simulations.

In the following section a semi-empirical framework for the description of differential constitutive equations for the Cauchy stress tensor is presented. When

J.F. Dijksman and G.D.C. Kuiken (eds.),
IUTAM Symposium on Numerical Simulation of Non-isothermal Flow of Viscoelastic Liquids, 21-35.

application of constitutive equation is met in numerical simulations of unsteady, nonisothermal flows in complex geometries, they should not be formulated in terms of molecular quantities, i.e. variables that represent aspects of the microstructure in its full detail. Consequently, in the presented framework only the foundation in the microstructure and its evolution in time are kept in a elementary way, while mathematical convenience is strived for. The framework used meant to include all existing models and illustrate in comprehensive way where they differ, resemble or can be modified and/or extended. The need for such modification and/or extension of existing models has been known for quite some time and is recently once more demonstrated in a paper of Baaijens et al. [2].

2. Constitutive Equations

The structure of a fluid may consist of, for example, (parts of) flexible macromolecules, rodlike molecules or suspended interacting flexible or rigid particles. This structure is modelled with "connector" vectors $\vec{R}_i$. The vectors $\vec{R}_i$ are said to represent so called structural elements. Different types of structural elements can be present in a fluid and the subscript "i" refers to elements of type "i". Differences find there origin not only in the behaviour of the elements themselves, but also in different interactions of the elements with their surroundings.

From the principle of virtual work it follows that the contribution to the Cauchy stress tensor of elements of type "i" is given by the Kramers expression for the average stress tensor [5, 12]:

$$\boldsymbol{\sigma}_i = \xi\nu < \vec{f}_i\vec{R}_i > \tag{1}$$

where ν is the number density of particles, $\vec{f}_i$ the connector force and $< \cdot >$ the average over the non-equilibrium configuration distribution function $\Psi(\vec{R}_i)$ of the vectors $\vec{R}_i$ (an average in the $\vec{R}$ − *space*). The factor ξ is a force reducing factor, introduced here to capture also models like Larsons partial extending model. The connector force $\vec{f}_i$ is considered to be a nonlinear function of both the length of the element vector $|\vec{R}|$ and the temperature T and can be derived from a potential of free elastic energy F_s:

$$\vec{f}_i = \frac{\partial F_s(|\vec{R}_i|, T)}{\partial \vec{R}_i} = c(|\vec{R}_i|, T)\vec{R}_i \tag{2}$$

The stress tensor is now written as:

$$\boldsymbol{\sigma}_i = \xi\nu < c(|\vec{R}_i|, T)\vec{R}_i\vec{R}_i > \tag{3}$$

For convenience the subscript i and the force reducing factor ξ will be omitted in this section.

The equation of motion for a vector $\vec{R}$ is postulated as:

$$\dot{\vec{R}} = \mathbf{L}\cdot\vec{R} - \mathbf{A}\cdot\vec{R} \tag{4}$$

with $\mathbf{L}$ the macroscopic velocity gradient tensor. The second order tensor $\mathbf{A}$ is a yet to be specified function of averaged, thus macroscopic variables, i.e. the stress, strain or strain rate. The level of description, a vector representing a force carrying element on the micro level, is taken in an averaged sense: $\dot{\vec{R}}$ does not include the actual rapid fluctuations of $\vec{R}$ due to thermal motion. The term $-\mathbf{A}\cdot\vec{R}$ represents the slippage of the element with respect to the continuum. $\mathbf{A}$ is called the slip tensor. In case a structural element deforms affine with the macroscopic deformation, $\mathbf{A} = 0$. The above equation is postulated for any deformation and not only for the special case of sudden, steplike deformations with no time available for relaxation. This implies that the equation of motion for $\vec{R}$ incorporates both convection and diffusion of $\vec{R}$ [9].

When the slip tensor $\mathbf{A}$ is allowed to be a function of micro structural variables such as the structural element vector $\vec{R}$ and the distribution function $\Psi(\vec{R})$, expression (4) has the same structure as a Langevin equation for a dumbbell [12]. It is normal practice to substitute the Langevin equation in the probability balance equation, resulting in the Smoluchowski equation, a convection diffusion equation for the probability function Ψ in the $\vec{R}$-space. Only for special cases, such as the Rouse model, the constitutive equation for the stress tensor $\boldsymbol{\sigma}$ can be obtained from the Smoluchowski equation [12]. Using approximations, also for other models, like the Larson partial extending model, a constitutive equation for $\boldsymbol{\sigma}$ can be found [10]. By choosing $\mathbf{A}$ to be a function of averaged variables only, a direct connection is made between the micro and macro level. With this starting supposition and the following scheme for deriving the constitutive equation for $\boldsymbol{\sigma}$, see below, solving the Smoluchowski equation is no longer needed.

The time derivative of the stress tensor (3) is given by:

$$\dot{\boldsymbol{\sigma}} = \nu(< c\dot{\vec{R}}\vec{R} > + < c\vec{R}\dot{\vec{R}} > + < \frac{\partial c}{\partial|\vec{R}|}\frac{d|\vec{R}|}{dt}\vec{R}\vec{R} > + < \frac{\partial c}{\partial T}\frac{dT}{dt}\vec{R}\vec{R} > \qquad (5)$$

The following approximations are introduced:

$$< \frac{\partial c}{\partial|\vec{R}|}\frac{d|\vec{R}|}{dt}\vec{R}\vec{R} > \approx < \frac{1}{c}\frac{\partial c}{\partial|\vec{R}|}\frac{d|\vec{R}|}{dt} > < c\vec{R}\vec{R} > \qquad (6)$$

$$< \frac{\partial c}{\partial T}\frac{dT}{dt}\vec{R}\vec{R} > \approx \dot{T} < \frac{\partial \ln c}{\partial T} > < c\vec{R}\vec{R} > \qquad (7)$$

With the expression (4) for $\dot{\vec{R}}$ and $|\vec{R}| = \sqrt{\vec{R}\cdot\vec{R}}$ the term $< \frac{1}{c}\frac{\partial c}{\partial|\vec{R}|}\frac{d|\vec{R}|}{dt} >$ in the first approximation can be worked out as:

$$< \frac{1}{c}\frac{\partial c}{\partial|\vec{R}|}\frac{d|\vec{R}|}{dt} > = K\mathbf{I} : (\mathbf{L}\cdot\boldsymbol{\sigma} + \boldsymbol{\sigma}\cdot\mathbf{L} - \mathbf{A}\cdot\boldsymbol{\sigma} - \boldsymbol{\sigma}\cdot\mathbf{A}) \qquad (8)$$

where the nonlinear material function K is defined by:

$$K = \frac{1}{\nu} < \frac{1}{c^2}\frac{1}{2|\vec{R}|}\frac{\partial c}{\partial|\vec{R}|} > \qquad (9)$$

When the expression between the brackets is constant, and thus $K = constant$, integration and using expression (2) yields the Warner force law, often called FENE (Finite Extensible Nonlinear Elastic) spring:

$$\vec{f} = \frac{1}{b - \nu K|\vec{R}|^2}\vec{R} \tag{10}$$

where b is an integration constant. The Warner force law is a good approximation of the inverse Langevin function; a complex, implicit force law that can be derived from molecular arguments [5]. Defining the material function H_T as:

$$H_T =< \frac{\partial \ln c}{\partial T} >, \tag{11}$$

using the shorthand notation:

$$\mathbf{A}_k = \mathbf{A} - K\boldsymbol{\sigma} : (\mathbf{D} - \mathbf{A})\mathbf{I} \tag{12}$$

and the definition of the upper convected derivative, the constitutive equation for $\boldsymbol{\sigma}$ is given by:

$$\overset{\nabla}{\boldsymbol{\sigma}} + \mathbf{A}_k \cdot \boldsymbol{\sigma} + \boldsymbol{\sigma} \cdot \mathbf{A}_k - \dot{T} H_T \boldsymbol{\sigma} = \mathbf{0} \tag{13}$$

For isothermal processes the last term on the left side of (13) equals zero. This demonstrates that for nonisothermal systems material parameters such as a time constant cannot simply be replaced by a temperature dependent function [5].

Stress Relaxation

There are constraints for the slip tensor **A**. For example, **A** should be chosen in a way that stress relaxation is properly described. Restricting for convenience to isothermal problems ($\dot{T} = 0$) and linear behaviour of the connector force ($\vec{f} = c\vec{R}$), then, after cessation of a flow ($\mathbf{L} = \mathbf{0}$) the constitutive equation (13) reads:

$$\dot{\boldsymbol{\sigma}} + \mathbf{A}(\mathbf{L} = \mathbf{0}) \cdot \boldsymbol{\sigma} + \boldsymbol{\sigma} \cdot \mathbf{A}(\mathbf{L} = \mathbf{0}) = \mathbf{0} \tag{14}$$

For $t \to \infty$, equilibrium should be obtained and thus $\dot{\boldsymbol{\sigma}} \to \mathbf{0}$. Also $\boldsymbol{\sigma} \to G\mathbf{I}, (G = \frac{1}{3}c|\vec{R}_0|^2)$, and therefore $\mathbf{A} \to \mathbf{0}$. Restricting **A** to be a function of $\boldsymbol{\sigma}$ and **L**, hence $\mathbf{A} = \mathbf{A}(\boldsymbol{\sigma}, \mathbf{L})$, it follows:

$$\mathbf{A} = \mathbf{A}(\boldsymbol{\sigma}, \mathbf{0}) \;= \mathbf{A}^* \to \mathbf{0} \;\; for \;\; t \to \infty \tag{15}$$

A rather general choice for $\mathbf{A}^*$ which satisfies this requirement is:

$$\begin{aligned} \mathbf{A}^* &= a_1\boldsymbol{\sigma} + a_2\boldsymbol{\sigma}^{-1} + a_3\mathbf{I} \\ with\, a_3 &\to -(Ga_1 + \frac{a_2}{G}) \;\; for \;\; t \to \infty \end{aligned} \tag{16}$$

where $a_i, (i = 1, 2, 3)$ are scalar functions of the invariants of $\boldsymbol{\sigma}$. In case $\mathbf{A} = \mathbf{A}^*$ the constitutive equation reads:

$$\overset{\nabla}{\boldsymbol{\sigma}} + 2a_1\boldsymbol{\sigma} \cdot \boldsymbol{\sigma} + 2a_3\boldsymbol{\sigma} + 2a_2\mathbf{I} = \mathbf{0} \tag{17}$$

Notice that this form of the constitutive equation, based on simple notions on stress relaxation, already comprises the rather complex models of Leonov and Giesekus. For:

$$a_1 = \frac{1}{4\lambda G} \quad a_2 = -\frac{G}{4\lambda} \quad a_3 = \frac{I_1 - G^2 I_1^-}{12\lambda G} \tag{18}$$

where $I_1 = tr(\boldsymbol{\sigma}), I_1^- = tr(\boldsymbol{\sigma}^{-1})$, the Leonov model is obtained:

$$\overset{\nabla}{\boldsymbol{\sigma}} + \frac{1}{2G\lambda}[\boldsymbol{\sigma} \cdot \boldsymbol{\sigma} - G^2\mathbf{I} - \frac{1}{3}(I_1 - G^2 I_1^-)\boldsymbol{\sigma}] = \mathbf{0} \tag{19}$$

The *elastic strain tensor* $\mathbf{B}_e$ of Leonov equals the tensor $< \vec{R}\vec{R} >$ as used here; the *irreversible rate of strain* $\mathbf{D}_p$ in the model of Leonov corresponds to the slip tensor $\mathbf{A}$. The special choice for the constitutive equation for $\mathbf{D}_p$ as proposed by Leonov, in literature often mentioned as seemingly arbitrary, provides a correct and quite general description of the phenomenon of stress relaxation. Another special property of the Leonov model is the choice of $\mathbf{A}$ as such that the contribution from *inter-element entropy* to the free energy is zero, see Section 4, equation (88).

Examples

For a number of well-know constitutive equations the slip tensors $\mathbf{A}$ are listed below. For these models the nonlinear material function (9) $K = 0$. This list is certainly not complete and can easily be extended.

Classical Rubber Elasticity (20)

$$\mathbf{A} = \mathbf{0} \tag{20}$$

Based on the Gordon-Schowalter convected derivative with a slip parameter ξ equation (21 includes the following models.

Johnson/Segalman, Upper Convected Maxwell ($\xi = 0$), Lower Convected Maxwell ($\xi = 2$), Corotational Maxwell ($\xi = 1$)

$$\mathbf{A} = \xi\mathbf{D} + \frac{1}{2\lambda}\mathbf{I} - \frac{G}{2\lambda}\boldsymbol{\sigma}^{-1} \tag{21}$$

Based on the partial extending convective derivative with slip parameter ξ equation (22) includes the following models.

Doi-Edwards Differential Approximation (23), Larson Partial Extending[1] (24), Larson Simplified Partial Extending (25), Larson Irreversible Partial Extending (26)

[1] The constitutive equation as Larson derived in ([10]) is erroneous. He missed one term with the integration of equation (31) in that paper. This term, $6G\xi(1-\xi) < \vec{n}\vec{n} : \mathbf{D} >< \vec{n}\vec{n} >$, should be added to the left side of his equation (32). Proceeding the deriviation of the model with the correct equation leads via a coupled two mode model to the constitutive equations as given here (24), [15].

$$\mathbf{A} = f(\boldsymbol{\sigma},\mathbf{D})\mathbf{I} + \frac{1}{2\lambda}\mathbf{I} - \frac{G}{2\lambda}\boldsymbol{\sigma}^{-1} \tag{22}$$

$$f(\boldsymbol{\sigma}) = \frac{1}{3G}(\boldsymbol{\sigma} : \mathbf{D}) \tag{23}$$

$$f(\boldsymbol{\sigma}) = \frac{\xi}{tr(\boldsymbol{\sigma})}[1 + \frac{6G(1-\xi)}{tr(\boldsymbol{\sigma})}](\boldsymbol{\sigma} : \mathbf{D}) \tag{24}$$

$$f(\boldsymbol{\sigma}) = \frac{\xi}{3G}[1 + \frac{2(1-\xi)}{3G}](\boldsymbol{\sigma} : \mathbf{D}) \tag{25}$$

$$f(\boldsymbol{\sigma}) = \frac{\xi}{tr(\boldsymbol{\sigma})}[1 + \frac{6G(1-\xi)}{tr(\boldsymbol{\sigma})}][\boldsymbol{\sigma} : \mathbf{D}]_+ \tag{26}$$

with in the last expression:

$$\begin{aligned} [\boldsymbol{\sigma} : \mathbf{D}]_+ &= \boldsymbol{\sigma} : \mathbf{D} \; for \; \boldsymbol{\sigma} : \mathbf{D} > 0 \\ [\boldsymbol{\sigma} : \mathbf{D}]_+ &= 0 \; for \; \boldsymbol{\sigma} : \mathbf{D} \leq 0 \end{aligned} \tag{27}$$

Leonov (28), Giesekus (29)

$$\mathbf{A} = \frac{1}{4G\lambda}\boldsymbol{\sigma} - \frac{G}{4\lambda}\boldsymbol{\sigma}^{-1} - \frac{tr(\sigma) - G^2 tr(\sigma^{-}1)}{12\lambda G}\mathbf{I} \tag{28}$$

$$\mathbf{A} = \frac{\alpha}{2\lambda G}\boldsymbol{\sigma} + \frac{1}{2\lambda}(1-2\alpha)\mathbf{I} - \frac{G}{2\lambda}(1-\alpha)\boldsymbol{\sigma}^{-1} \tag{29}$$

The following models have a stress or strain rate dependent characteristic material time functions. Sometimes, the Gordon-Schowalter convective derivative is used.

Phan-Thien/Tanner a and b (31, 32), White/Metzner (33 and $\xi = 0$)

$$\mathbf{A} = \xi\mathbf{D} + \frac{1}{2\lambda}\mathbf{I} - \frac{G}{2\lambda}\boldsymbol{\sigma}^{-1} \tag{30}$$

$$\lambda = \lambda_0 \exp[-\frac{\varepsilon}{G}tr(\boldsymbol{\sigma} - G\mathbf{I})] \tag{31}$$

$$\lambda = \lambda_0[1 + \frac{\varepsilon}{G}tr(\boldsymbol{\sigma} - G\mathbf{I})]^{-1} \tag{32}$$

$$\lambda = \lambda_0[1 + \alpha\lambda_0\sqrt{II_D}]^{-1} \tag{33}$$

It is possible to construct a class of FENE models by combining different material functions K (9) with different slip tensors **A** and different functions for the characteristic material time λ [14].

3. Thermodynamics

The equation of conservation of energy, also known as the first law of thermodynamics, in its local form is given by:

$$\rho \dot{u} = -\vec{\nabla} \cdot \vec{q} + \boldsymbol{\sigma} : \mathbf{D} + \rho r \tag{34}$$

The internal heat production other than viscous dissipation as represented by the term ρr will be ignored for simplicity.
According to the second law of thermodynamics, also referred to as the Clausius-Duhem inequality, the entropy production in a thermomechanical process will be equal to or greater than zero:

$$\rho T \dot{s} - \rho \dot{u} + \boldsymbol{\sigma} : \mathbf{D} - \frac{\vec{q} \cdot \vec{\nabla} T}{T} \geq 0 \tag{35}$$

The following variables are chosen as the independent state variables for a viscoelastic fluid:

$$\rho, \mathbf{L}, \mathbf{B}_{e_i}, T, \text{ and } \vec{\nabla} T \tag{36}$$

where $\mathbf{B}_{e_i}$ is a shorthand notation for the elastic strain tensor $< \vec{R}_i \vec{R}_i >$. The constitutive quantities of interest are:

$$\boldsymbol{\sigma}, \vec{q}, u \text{ and } s \tag{37}$$

The Entropy Inequality
The principle of equipresence states that any of these quantities is a function of all independent state variables, unless proven otherwise:

$$\boldsymbol{\sigma} = \boldsymbol{\sigma}(\rho, \mathbf{L}, \mathbf{B}_{e_i}, T, \vec{\nabla} T) \tag{38}$$
$$\vec{q} = \vec{q}(\rho, \mathbf{L}, \mathbf{B}_{e_i}, T, \vec{\nabla} T) \tag{39}$$
$$u = u(\rho, \mathbf{L}, \mathbf{B}_{e_i}, T, \vec{\nabla} T) \tag{40}$$
$$s = s(\rho, \mathbf{L}, \mathbf{B}_{e_i}, T, \vec{\nabla} T) \tag{41}$$

The Helmholtz free energy f is defined as:

$$f \equiv u - Ts \tag{42}$$

As a consequence of this equation and, again, the principle of equipresence, also f can be a function of all state variables:

$$f = f(\rho, \mathbf{L}, \mathbf{B}_{e_i}, T, \vec{\nabla} T) \tag{43}$$

Substituting equations (42) and (43) in equation (35) yields:

$$-\rho\left(\frac{\partial f}{\partial \rho}\dot{\rho}+\frac{\partial f}{\partial \mathbf{L}}:\dot{\mathbf{L}}+\sum_i\left(\frac{\partial f}{\partial \mathbf{B}_{e_i}}\right)^c:\dot{\mathbf{B}}_{e_i}+\frac{\partial f}{\partial T}\dot{T}+\frac{\partial f}{\partial \vec{\nabla} T}\cdot\dot{\vec{\nabla} T}+\dot{T}s\right)$$
$$+\boldsymbol{\sigma}:\mathbf{D}-\frac{\vec{q}\cdot\vec{\nabla}T}{T}\geq 0 \quad (44)$$

in which $\dot{\rho}$, $\dot{\mathbf{L}}$, $\dot{\mathbf{B}}_{e_i}$, $\dot{T}$ and $\dot{\vec{\nabla} T}$ can be varied independently. This implies that:

$$\frac{\partial f}{\partial \mathbf{L}}=\mathbf{0},\ \frac{\partial f}{\partial T}=-s \text{ and } \frac{\partial f}{\partial \vec{\nabla} T}=\vec{0} \quad (45)$$

and thus, using also the definition (42) of the free energy, it is seen that:

$$f = f(\rho, \mathbf{B}_{e_i}, T) \quad (46)$$
$$u = u(\rho, \mathbf{B}_{e_i}, T) \quad (47)$$
$$s = s(\rho, \mathbf{B}_{e_i}, T) \quad (48)$$

For $\dot{\mathbf{B}}_{e_i}$ the following expression is derived using the equation of motion (4) for $\vec{R}_i$:

$$\dot{\mathbf{B}}_{e_i}=(\mathbf{L}-\mathbf{A}_i)\cdot\mathbf{B}_{e_i}+\mathbf{B}_{e_i}\cdot(\mathbf{L}^c-\mathbf{A}_i^c) \quad (49)$$

This enables us to rewrite the third term in equation (44) as:

$$\sum_i\left(\frac{\partial f}{\partial \mathbf{B}_{e_i}}\right)^c:\dot{\mathbf{B}}_{e_i}=2\sum_i\mathbf{B}_{e_i}\cdot\left(\frac{\partial f}{\partial \mathbf{B}_{e_i}}\right)^c:(\mathbf{D}-\mathbf{A}_i) \quad (50)$$

With this result and equation (45), equation (44) can be recast into:

$$-\rho\left(\frac{\partial f}{\partial \rho}\dot{\rho}+2\sum_i\mathbf{B}_{e_i}\cdot\left(\frac{\partial f}{\partial \mathbf{B}_{e_i}}\right)^c:(\mathbf{D}-\mathbf{A}_i)\right)+\boldsymbol{\sigma}:\mathbf{D}-\frac{\vec{q}\cdot\vec{\nabla}T}{T}\geq 0 \quad (51)$$

The stress tensor is assumed to be the sum of a number of elastic parts $\boldsymbol{\sigma}_{e_i}$, which are related to the strain tensors $\mathbf{B}_{e_i}$, and a purely viscous part $\boldsymbol{\sigma}_v$:

$$\boldsymbol{\sigma}=\sum_i\boldsymbol{\sigma}_{e_i}+\boldsymbol{\sigma}_v \quad (52)$$

Using this, expression (51) can be rewritten in:

$$-\rho\left(\frac{\partial f}{\partial \rho}\dot{\rho}+2\sum_i\left(\mathbf{B}_{e_i}\cdot\left(\frac{\partial f}{\partial \mathbf{B}_{e_i}}\right)^c-\frac{1}{\rho}\boldsymbol{\sigma}_{e_i}\right):\mathbf{D}-\sum_i\mathbf{B}_{e_i}\cdot\left(\frac{\partial f}{\partial \mathbf{B}_{e_i}}\right)^c:\mathbf{A}_i\right)$$
$$+\boldsymbol{\sigma}_v:\mathbf{D}-\frac{\vec{q}\cdot\vec{\nabla}T}{T}\geq 0 \quad (53)$$

The elastic parts of the stress are defined as:

$$\boldsymbol{\sigma}_{e_i} = 2\rho\mathbf{B}_{e_i} \cdot \left(\frac{\partial f}{\partial \mathbf{B}_{e_i}}\right)^c \tag{54}$$

Thus the free energy is a potential for the stress tensors $\boldsymbol{\sigma}_{e_i}$. Restricting to incompressible materials, the Clausius-Duhem inequality (53) becomes:

$$\boldsymbol{\sigma}_v : \mathbf{D} + \sum_i \boldsymbol{\sigma}_{ei} : \mathbf{A}_i - \frac{\vec{q} \cdot \vec{\nabla} T}{T} \geq 0 \tag{55}$$

The rate of thermal entropy production $\delta_T$is defined as:

$$\delta_T = -\frac{\vec{q} \cdot \vec{\nabla} T}{T} \tag{56}$$

while the rate of mechanical entropy production δ_m is defined as:

$$\delta_m = \boldsymbol{\sigma}_v : \mathbf{D} + \sum_i \boldsymbol{\sigma}_{ei} : \mathbf{A}_i \tag{57}$$

Thus equation (55) can be reduced to:

$$\delta_m + \delta_T \geq 0 \tag{58}$$

which expresses that the sum of the mechanical and the thermal entropy production is equal to or greater than zero. However, these production rates do *not* have to be separately non-negative.

For an incompressible fluid it follows from (48):

$$f = f(\mathbf{B}_{e_i}, T) \tag{59}$$
$$u = u(\mathbf{B}_{e_i}, T) \tag{60}$$
$$s = s(\mathbf{B}_{e_i}, T) \tag{61}$$

Using this, the original inequalty (35) turns into:

$$\rho\left(T\frac{\partial s}{\partial T} - \frac{\partial u}{\partial T}\right)\dot{T} + \boldsymbol{\sigma} : \mathbf{D} - \rho\dot{u}|_T + \rho T\dot{s}|_T - \frac{\vec{q} \cdot \vec{\nabla} T}{T} \geq 0 \tag{62}$$

Variation of $\dot{T}$ leads to:

$$T\frac{\partial s}{\partial T} = \frac{\partial u}{\partial T} \tag{63}$$

$$\boldsymbol{\sigma} : \mathbf{D} - \rho\dot{u}|_T + \rho T\dot{s}|_T - \frac{\vec{q} \cdot \vec{\nabla} T}{T} \geq 0 \tag{64}$$

The rate of mechanical entropy production δ_m is now given by:

$$\delta_m = \boldsymbol{\sigma} : \mathbf{D} - \rho\dot{u}|_T + \rho T\dot{s}|_T \tag{65}$$

With definition of the free energy (42) and comparing (57) with (65) one gets:

$$\rho\dot{f}|_T = \sum_i \boldsymbol{\sigma}_{e_i} : (\mathbf{D} - \mathbf{A}_i) \tag{66}$$

The Energy Equation

Next the equation of conservation of energy (34) is dealt with. From the definition of the free energy (42):

$$\dot{u} = \dot{f} + T\dot{s} + \dot{T}s \tag{67}$$

Using this relation (67) and the expressions $(45)^b$ for s, (49) for $\dot{\mathbf{B}}_{e_i}$ and (59) for f, the energy equation (34) is rewritten to:

$$\rho c_v \dot{T} = -\vec{\nabla}\cdot\vec{q} + \boldsymbol{\sigma} : \mathbf{D} - \sum_i \boldsymbol{\sigma}_{e_i} : (\mathbf{D} - \mathbf{A}_i) + \rho T \sum_i \frac{\partial \frac{1}{\rho}\boldsymbol{\sigma}_{e_i}}{\partial T} : (\mathbf{D} - \mathbf{A}_i) \tag{68}$$

where the specific heat c_v at constant volume is defined as:

$$c_v = -T\frac{\partial^2 f}{\partial T^2} = \frac{\partial u}{\partial T} \tag{69}$$

The form of the energy equation that is often used in literature [3, 13] differs from (68); the last two terms of (68) are not incorporated. With the partitioning of $\boldsymbol{\sigma}$ as given in (52), equation (68) can also be written in a form where all terms are expressed in the different stress parts:

$$\rho c_v \dot{T} = -\vec{\nabla}\cdot\vec{q} + \boldsymbol{\sigma}_v : \mathbf{D} + \sum_i \boldsymbol{\sigma}_{e_i} : \mathbf{A}_i + \rho T \sum_i \frac{\partial \frac{1}{\rho}\boldsymbol{\sigma}_{e_i}}{\partial T} : (\mathbf{D} - \mathbf{A}_i) \tag{70}$$

In order to analyse the seperate contributions from entropy and energy elasticity, the energy equation is written in a different way starting with the internal energy as a function of the strain and temperature (60). In the case of pure entropy elasticity the internal energy is a function of the temperature only, $u = u(T)$. Using the definition of specific heat c_v (69) and the expression for $\dot{\mathbf{B}}_{ei}$ (49), the energy equation is written as:

$$\rho c_v \dot{T} = -\vec{\nabla}\cdot\vec{q} + \boldsymbol{\sigma} : \mathbf{D} - 2\rho \sum_i \mathbf{B}_{e_i} \cdot \left(\frac{\partial u}{\partial \mathbf{B}_{e_i}}\right)^c : (\mathbf{D} - \mathbf{A}_i) \tag{71}$$

The last term in this equation represents the mechanical energy supply related to the energy elasticity. With the definition (54) of the elastic stresses $\boldsymbol{\sigma}_{ei}$, the definition (42) of the free energy f and the relation $(45)^b$ between the specific entropy s and the free energy f, one gets:

$$\begin{aligned} \boldsymbol{\sigma}_{e_i} &= 2\rho\left(\frac{\partial u}{\partial \mathbf{B}_{e_i}}\right)^c \cdot \mathbf{B}_{e_i} + 2\rho T \mathbf{B}_{e_i} \cdot \left(\frac{\partial^2 f}{\partial \mathbf{B}_{e_i}\partial T}\right)^c \\ &= 2\rho\left(\frac{\partial u}{\partial \mathbf{B}_{e_i}}\right)^c \cdot \mathbf{B}_{e_i} + \rho T \frac{\partial \frac{1}{\rho}\boldsymbol{\sigma}_{e_i}}{\partial T} \end{aligned} \tag{72}$$

With this result the energy equation (71) is rewritten to:

$$\rho c_v \dot{T} = -\vec{\nabla} \cdot \vec{q} + \boldsymbol{\sigma} : \mathbf{D} - \sum_i (\boldsymbol{\sigma}_{ei} - \rho T \frac{\partial \frac{1}{\rho} \boldsymbol{\sigma}_{e_i}}{\partial T}) : (\mathbf{D} - \mathbf{A}_i) \tag{73}$$

or, with the definitions (3) and (11) of the stress tensor $\boldsymbol{\sigma}_{e_i}$ and the material function H_T respectively and using the approximation (7) this equation finally becomes:

$$\rho c_v \dot{T} = -\vec{\nabla} \cdot \vec{q} + \boldsymbol{\sigma} : \mathbf{D} - \sum_i (1 - TH_{T_i} + \frac{T}{\rho} \frac{\partial \rho}{\partial T}) \boldsymbol{\sigma}_{e_i} : (\mathbf{D} - \mathbf{A}_i) \tag{74}$$

The material function H_T appears in the constitutive equation (13) and in the energy equation (74). Knowledge of this function is therefore a key to correct thermorheological modelling.

It is seen that the partial derivatives of the elastic stresses with respect to the temperature, also expressible in terms of the material function H_T, rule the partition between entropic and energy elasticity. One can observe immediately that in case the elastic stress $\boldsymbol{\sigma}_{ei}$ is a linear function of the temperature, the energy elasticity is zero. Linear dependency of the stress on temperature is a well known result for polymer chains that are modelled as a *freely-jointed chain* for which the internal energy doesn't change on loading; at least for not to large extensions of the chains. With the stress according to (52), equation (70) is regained.

Examples

Braun [8] modeled the thermorheological behaviour of viscoelastic fluids based on a group of constitutive equations of the differential type and linear relation between $\boldsymbol{\sigma}_e$ and $\mathbf{B}_e$:

$$\boldsymbol{\sigma}_{e_i} = G(T)\mathbf{B}_e \tag{75}$$

The temperature dependence of the modulus G is described by:

$$G = G_0 \bar{T}^{\alpha} \text{ ; with } G_0 = G(T_0) \text{ and } \bar{T} = \frac{T}{T_0} \tag{76}$$

where T_0 is the reference temperature and α is a model parameter, that can be either positive or negative. In his paper [8], Braun considered the influence of the parameter α on the thermorheological behaviour of a viscoelastic fluid using a stress-growth experiment with relaxation. He showed that the parameter α determines whether the process is exo-, endo- or isothermal. For entirely entropy elastic fluids ($\alpha = 1$), the experiment is exothermal until relaxation, which is isothermal. For entirely energy elastic fluids ($\alpha = 0$), the entire experiment is exothermal.

Sarti and Esposito [16] have found isothermal relaxation (entropy elastic behaviour) for different polyisobutylenes and endothermal relaxation of polyvinylacetate at $20°C$ above its glass transition temperature. In terms of relation (76) this can be explained by assuming $\alpha > 1$.

For an adiabatic steplike deformation process, the Larson Partial Extending model (24) gives an entropy production rate:

$$\delta_m = \boldsymbol{\sigma}_e : \mathbf{A} = \xi[1 + \frac{6G(1-\xi)}{tr(\boldsymbol{\sigma})}](\boldsymbol{\sigma}_e : \mathbf{D}) \tag{77}$$

Reversing the step, i.e. reversing the sign of $\boldsymbol{D}$, leads to a negative entropy production. The irreversible Larson model (26) does not suffer from this problem. The same problem as sketched for the Larson Partial Extending Model is found in constitutive models with the Gordon-Schowalter convective derivative if used for compressible fluids.

The stress tensor in the Doi/Edwards model is given by [12]:

$$\boldsymbol{\sigma} = 3\nu kT < \vec{n}\vec{n} >; \quad \vec{n} = \frac{\vec{R}}{|\vec{R}|} \tag{78}$$

If the pure viscous stress $\boldsymbol{\sigma}_v = \mathbf{0}$, then, using the slip tensor for the differential approximation of the Doi/Edwards model (23), the mechanical entropy production rate (57) is found to be:

$$\delta_m = \boldsymbol{\sigma}_e : \mathbf{A} = \boldsymbol{\sigma}_e : \mathbf{D} \tag{79}$$

and thus the mechanical supplied energy is completely transformed into entropy production. Also the free elastic energy rate at constant temperature (66) is always zero, suggesting that no elastic energy can be stored. What is wrong? In the next section (4) this problem is dealt with. The Helmholtz free energy derived from micro structural arguments shows an extra term compared with expression (66). This extra term, related to inter-element entropy, did not appear from the thermodynamical analysis in this section. Although this term might be incorporated, the framework of thermodynamics does not indicate when and how this should be done. It is, consequently, necessary to take the microstructure into consideration in order to deal with this problem. In the next section it is shown that for incompressible fluids ($tr(\mathbf{D}) = 0$) the inter-element entropy only contributes to the free energy rate when $tr(\mathbf{A}) \neq 0$. A constitutive model with $tr(\mathbf{A}) = 0$ is the Leonov model.

4. The Free energy derived from the microstructure

In general, structures in fluids will contribute to the free energy in two different ways. The first orginates from the individual structural elements and is called here the element free elastic energy. It is written as a potential $F_s(|\vec{R}|)_i$. There are, again, two special cases for this free elastic energy: entropy elasticity relating to the internal entropy of the structural element and the (internal) energy elasticity relating to the internal energy of the structural element. If the free energy of an element is noted as $F_s(|\vec{R}|, T) = U_s(|\vec{R}|, T) - TS_s(|\vec{R}|, T)$, the specific free energy is given by:

$$\begin{aligned} f_s &= \nu < F_s(|\vec{R}|, T) > + constant \\ &= \nu(< U_s(|\vec{R}|, T) > - T < S_s(|\vec{R}|, T) >) + constant \end{aligned} \tag{80}$$

It is normal practice to choose the constant such that $f_s = 0$ at equilibrium.
The second contribution to the free elastic energy is of a combinatoric nature. It is arising from the number of ways a set of N elements can be chosen to satisfy the distribution function Ψ and it thus relates to the inter-element entropy and is given by [9]:

$$f_c = \nu kT < \log \frac{\Psi}{\Psi_0} > \tag{81}$$

The integration constant is chosen zero such that again$f_c = 0$ at equilibrium.
From these results the free elastic energy rate at constant temperature is obtained:

$$\begin{aligned} \dot{f}|_T &= \dot{f}_s|_T + \dot{f}_c|_T \\ &= \nu(< \frac{\partial F_s(|\vec{R}|, T)}{\partial \vec{R}} \cdot \dot{\vec{R}} > + kT < \overset{\cdot}{\log \frac{\Psi}{\Psi_0}} >) \end{aligned} \tag{82}$$

This result should be written in terms of the slip tensor $\mathbf{A}$. Using (4), the first term in (82) is easily rewritten to:

$$\dot{f}_s|_T = \nu < c(|\vec{R}|, T)\vec{R}\vec{R} >: (\mathbf{L} - \mathbf{A}) \tag{83}$$

The last term in (82) can be worked out by introducing the *configuration deformation tensor* $\mathbf{F_c}$:

$$\vec{R} = (\vec{\nabla}_{R_0}\vec{R})^c \cdot \vec{R}_0 = \mathbf{F}_c \cdot \vec{R}_0 \tag{84}$$

Comparing the time derivative $\dot{\vec{R}}$ taken from (84) with the equation of motion for the vector $\vec{R}$ leads to:

$$\dot{\mathbf{F}}_c \cdot \mathbf{F}_c^{-1} = (\vec{\nabla}_{R_0} \cdot \dot{\vec{R}})^c = \mathbf{L} - \mathbf{A} \tag{85}$$

With the probability balance equation [12]:

$$\dot{\Psi} + \Psi \vec{\nabla}_R \cdot \dot{\vec{R}} = 0 \tag{86}$$

it is found that:

$$\frac{\dot{\Psi}}{\Psi} = tr(\mathbf{L} - \mathbf{A}) \tag{87}$$

The second term in (82) is now written as:

$$\dot{f}_c|_T = -\nu kT tr(\mathbf{L} - \mathbf{A}) \tag{88}$$

Combining (83) and (88), the free elastic energy rate in terms of the slip tensor $\mathbf{A}$ is obtained:

$$\dot{f}|_T = \nu < c(|\vec{R}|, T)\vec{R}\vec{R} >: (\mathbf{D} - \mathbf{A}) - \nu kT tr(\mathbf{L} - \mathbf{A}) \tag{89}$$

Notice that incompressible fluids ($tr(\mathbf{D}) = 0$) only can have a contribution from the inter-element entropy to the free energy rate if $tr(\mathbf{A}) \neq 0$. The Leonov model is an example for which $tr(\mathbf{A}) = 0$.

A more complete modelling should incorporate the contribution of the inter-element entropy. In that case a coupled two mode constitutive model is derived [15] which, after some approximations can be transformed into the Larson Partial Extending model. The full thermorheological model for this two mode model is subject of current research.

5. Concluding Remarks

Based on the idea of nonaffine motion, a framework is given for differential constitutive equations. The framework allows for a systematic way of comparing different models and extending them. The first and second law of thermodynamics were expressed in terms of the framework. The partition of entropy and energy elasticity was elucidated and for some known models it was shown were they may cause problems or fail. It is demonstrated that when the free energy is derived from a model of the microstructure, although the model is not very detailed, an expression is found that has one extra term compared to the expression derived from general thermodynamics. This extra term is needed for a correct incorporation of the differential form of the Doi/Edwards constitutive model.

List of symbols

Symbols:

$\mathbf{A}$: slip tensor
$\mathbf{B}$: Finger tensor
c_v : specific heat at constant volume
$\mathbf{D}$: symmetric part of the rate of deformation tensor
F : elastic potential
$\mathbf{F}$: deformation tensor
$\vec{f}$: connector force
f : specific Helmholtz free energy
G : modulus
H_T : material function describing temperature dependency
$\mathbf{I}$: unit tensor
K : material function describing nonlinear extensibility
$\mathbf{L}$: rate of deformation tensor
$\vec{q}$: heat flux
$\vec{R}$: Vector representing a structural element
r : specific internal heat production
s : specific entropy
T : absolute temperature
u : specific internal energy

α : temperature dependence parameter

δ_m : mechanical energy dissipation (entropy production)
ρ : density
σ : Cauchy stress tensor
λ : relaxation time
ξ : slip factor
Ψ : configuration distribution function
$\vec{\nabla}$: vector differential operator

Indices:

$\mathbf{A_e}$: elastic part of $\mathbf{A}$
$\mathbf{A_p}$: plastic part of $\mathbf{A}$
$\overset{\triangledown}{\mathbf{A}}$: upper convective derivative of $\mathbf{A}$; $\overset{\triangledown}{\mathbf{A}} = \dot{\mathbf{A}} - \mathbf{L} \cdot \mathbf{A} - \mathbf{A} \cdot \mathbf{L}^c$

References

1. Astarita, G., and G.C. Sarti, "The dissipative mechanism in flowing polymers: theory and experiments", J. Non-Newtonian Fluid Mechanics, **1**: 39–50 (1976).
2. Baaijens, F.P.T, H.P.W. Baaijens, G.W.M. Peters and H.E.H. Meijer, "An experimental and numerical investigation of a viscoelastic flow around a cylinder", accepted for publication, J. of Rheology, (1993).
3. Baaijens, F.P.T, "Numerical Analysis of the Non-isothermal Viscoelastic Flow Around a Cilinder",IUTAM Symposium on Numerical Simulation of Nonisothermal Flow of Viscoelastic Liquids, 1-3 November 1993, Kerkrade, The Netherlands.
4. Bird, R.B., R.C. Armstrong, O. Hassager, "Dynamics of Polymeric Liquids", Vol 1, John Wiley & Sons, New York (1987).
5. Bird, R.B., C.F. Curtiss, R.C. Armstrong, O. Hassager, "Dynamics of Polymeric Liquids", Vol 2, John Wiley & Sons, New York (1987).
6. Braun, H., and Chr. Friedrich, "Transient processes in Couette flow of a Leonov fluid influenced by dissipation", J. of Non-Newtonian Fluid Mechanics, **33**: 39–51 (1989).
7. Braun, H., and Chr. Friedrich, "Dissipative behaviour of viscoelastic fluids derived from rheological constitutive equations", J. Non-Newtonian Fluid Mechanics, **38**: 81–91 (1990).
8. Braun, H., "A model for the thermorheological behavior of viscoelastic fluids", Rheologica Acta, **30**: 523–529 (1991).
9. Larson, R.G., "Convection and Diffusion of Polymer Network Strands", J. Non-Newtonian Fluid Mechanics, **13**, 279-308, (1983).
10. Larson, R.G., "A Constitutive Equation for Polymer Melts Based on Partially Extending Strand Convection", J. Rheology, **28**(5), 545-571, (1984).
11. Larson, R.G., J. Non-Newtonian Fluid Mechanics, **24**, 331-342, (1987).
12. Larson, R.G., "Constitutive Equations for Polymer Melts and Solutions", Butterworths, London (1988).
13. Luo, X.-L and R.I. Tanner, "A Pseudo-time Integral Method for Non-isothermal Viscoelastic Flows and its Application to Extrusion Simulation", Rheol. Acta 26: 499-507 (1987)
14. Peters, G.W.M., "Constitutive Models for the Stress Tensor Based on Nonaffine Motion", Faculty of Mechanical Engineering, Eindhoven University of Technology, The Netherlands, Internal Report, WFW93.140 (1993).
15. Peters, G.W.M., "The Constitutive Equation of Larson for the Stress Tensor Based on Nonaffine Motion", Faculty of Mechanical Engineering, Eindhoven University of Technology, The Netherlands, Internal Report, WFW93.141 (1993).
16. Sarti, G.C., and N. Esposito, "Testing thermodynamic constitutive equations for polymers by adiabatic deformation experiments", J. of Non-Newtonian Fluid Mechanics, **3**: 65–76 (1977/1978).

NUMERICAL SIMULATION OF A VISCOELASTIC FLUID WITH ANISOTROPIC HEAT CONDUCTION

P. WAPPEROM and M.A. HULSEN
Delft University of Technology, Laboratory for Aero and Hydrodynamics, Rotterdamseweg 145, 2628AL Delft, The Netherlands

Abstract. For the nonisothermal flow of a viscoelastic fluid we have taken into account temperature dependency of the relaxation times and the viscosities in the constitutive equation for the stress. In the energy equation the heat flux is specified by Fourier's law, where anisotropic heat conduction has been taken into account. Furthermore one has to specify which part of the stress work is dissipated and which part is stored as elastic energy. The equations are solved with a finite element method for the balance equations and a streamline integration method for the constitutive equation. The influence of the Deborah number, the Péclet number and the cooling temperature are examined in a flow through a 4 to 1 contraction.

Key words: shift factors, dissipation, anisotropic heat conduction, finite elements, streamline integration, axisymmetrical 4 to 1 contraction

1. Introduction

In the last 15 years much attention has been paid to the numerical simulation of isothermal flows of viscoelastic fluids. For recent reviews see Crochet (1989) and Keunings (1989). For steady problems the use of streamline integration methods made it possible to solve the equations for high Deborah numbers. Relatively small numbers of publications have appeared in on nonisothermal flows of viscoelastic fluids, especially not for high Deborah numbers. Therefore we extended the streamline integration method of Hulsen & van der Zanden (1991) to nonisothermal flows.

In section 2 of this article the general differential equations describing a nonisothermal viscoelastic flow are given. It starts with a short overview of the conservation laws we have used. Then we describe the constitutive equations for the stress and heat flux. For the stress equation we take into account that the viscosity and relaxation time depend on the temperature. The heat flux will be described with Fourier's law, where the heat conduction may be anisotropic. In the derivation of a temperature equation from the conservation of energy one has to specify which part of the stress work is dissipated and which part is stored as elastic energy. Section 3 describes the numerical implementation of the equations for nonisothermal, viscoelastic fluid flows. The finite element method (FEM) has been used to solve the momentum and temperature equations and a streamline integration method to solve the constitutive equation for the stress. Section 4 describes the boundary conditions of the 4 to 1 contraction and the characteristic quantities. The results of our calculations can be found in section 5. Finally some conclusions are given in section 6.

J.F. Dijksman and G.D.C. Kuiken (eds.),
IUTAM Symposium on Numerical Simulation of Non-isothermal Flow of Viscoelastic Liquids, 37-55.

2. Basic Equations

2.1. Conservation Laws

In this subsection we give a short description of the four balance equations we have used to describe the nonisothermal flow of a fluid. We will assume that the density of the fluid ρ is constant and that the flow is steady. Furthermore we will neglect body forces in the momentum equations and external heat sources in the energy equation. This leads to the following equations in a fixed bounded space Ω

– conservation of mass

$$\nabla \cdot \underline{v} = 0, \tag{1}$$

– balance of momentum

$$\rho \underline{v} \cdot \nabla \underline{v} + \nabla p = \nabla \cdot \underline{\underline{\tau}}, \tag{2}$$

in which $\underline{v}$ is the velocity, ρ the density, p the pressure and $\underline{\underline{\tau}}$ the extra-stress tensor. Note that the total stress $\underline{\underline{\sigma}}$ has been split into an isotropic part $p\underline{\underline{I}}$ and a deviatoric part $\underline{\underline{\tau}}$.

– balance of moment of momentum

$$\underline{\underline{\tau}} = \underline{\underline{\tau}}^T \tag{3}$$

– balance of energy

$$\rho \underline{v} \cdot \nabla u = \underline{\underline{\tau}} : \underline{\underline{d}} - \nabla \cdot \underline{\phi}_q, \tag{4}$$

where u is the internal energy and $\underline{\phi}_q$ the heat flux vector.

In these equations there are two quantities that must be specified by a constitutive equation, the stress and the heat flux. Furthermore we need an expression for the temperature instead of the energy, because some parameters depend on the temperature. These will be the topics of the section 2.2.

2.2. Constitutive Equations

2.2.1. Stress

The extra-stress tensor $\underline{\underline{\tau}}$ is determined by the deformation history of a fluid particle and has to be specified by a constitutive equation. For viscoelastic fluids it consists of a contribution of the Newtonian solvent and a contribution of the polymer part. To fit the behaviour of the polymeric part well, it is often necessary to consider multi-mode models:

$$\underline{\underline{\tau}} = 2\eta_s \underline{\underline{d}} + \sum_{k=1}^{K} \underline{\underline{\tau}}_k, \tag{5}$$

in which η_s is the solvent viscosity, K the number of modes and Euler's rate of deformation tensor $\underline{\underline{d}} = (\underline{\underline{L}} + \underline{\underline{L}}^T)/2$, with the velocity gradient $\underline{\underline{L}}^T = \nabla \underline{v}$. A substress $\underline{\underline{\tau}}_k$ can be specified by a differential or integral equation. In this article we will use a differential model, the Giesekus model. The differential equation will be given in terms of an internal deformation tensor $\underline{\underline{b}}_k$, as used in the derivation of the Giesekus model by Giesekus (1982). This is a more convenient form to derive an expression for the dissipation as described in section 2.2.3. Furthermore it will be used in the expression for the heat flux vector in section 2.2.2. It can be proved that $\underline{\underline{b}}_k$ is a positive definite tensor, see Hulsen (1990). For the Giesekus model the stress tensor $\underline{\underline{\tau}}_k$ can then be found from the internal deformation tensor $\underline{\underline{b}}_k$ by

$$\underline{\underline{\tau}}_k = \frac{\eta_k}{\lambda_k} \left(\underline{\underline{b}}_k - \underline{\underline{I}} \right), \tag{6}$$

where λ_k is the relaxation time and η_k the viscosity of mode k. In terms of the internal deformation tensor, the Giesekus model reads

$$\lambda_k \overset{\nabla}{\underline{\underline{b}}}_k + \left(\underline{\underline{I}} + \alpha_k \left(\underline{\underline{b}}_k - \underline{\underline{I}} \right) \right) \cdot \left(\underline{\underline{b}}_k - \underline{\underline{I}} \right) = \underline{\underline{0}}, \tag{7}$$

with the upper convected derivative defined as

$$\overset{\nabla}{(\ )} = \dot{(\ )} - \underline{\underline{L}} \cdot (\ ) - (\ ) \cdot \underline{\underline{L}}^T. \tag{8}$$

The constants $0 \leq \alpha_k < 1$ denote the amount of anisotropic drag. For $\alpha_k = 0$ the drag is isotropic and the model then reduces to the multi-mode upper-convected Maxwell model. The amount of anisotropy increases with increasing α_k.

The only parameters that are assumed temperature dependent are the viscosities and relaxation times. This temperature dependency is given by a shift factor a_T with which these parameters are multiplied. In our examples in section 5 there is no Newtonian viscosity. In principle the shift of the multi-mode viscosities and relaxation times can be chosen different for each mode:

$$\left. \begin{aligned} \log\left(\frac{\eta_k}{\eta_{k,\mathrm{ref}}}\right) &= \log a_{T,k}, \\ \log\left(\frac{\lambda_k}{\lambda_{k,\mathrm{ref}}}\right) &= \log a_{T,k}, \end{aligned} \right\} \quad k = 1, \ldots, K, \tag{9}$$

where the subscript $_{\mathrm{ref}}$ denotes a parameter at reference temperature. In section 5, however, we will use the same Andrade shift factor for each mode:

$$\log a_T = c_1 \left(\frac{1}{T} - \frac{1}{T_{\mathrm{ref}}} \right), \tag{10}$$

where c_1 is a shifting constant and T_{ref} the reference temperature.

2.2.2. Heat Flux

The heat flux vector in equation (4) has to be specified by a constitutive equation. A widely used constitutive equation is Fourier's law, which has the general form

$$\underline{\phi}_q = -\underline{\underline{\kappa}} \cdot \nabla T, \tag{11}$$

where the heat conduction tensor $\underline{\underline{\kappa}}$ can be a function of temperature and the internal deformation tensor. In this article we assume that it only depends on the internal deformation tensor. Mostly an isotropic heat conduction is assumed. Measurements of the heat conductivity by Hands (1980) and Washo & Hansen (1969), however, show that the increase in the direction of the orientation, and a corresponding decrease in the directions perpendicular to it, can be considerable.

For a one mode model the most general model that can be obtained for the heat conduction tensor, is

$$\underline{\underline{\kappa}} = \kappa_0 \underline{\underline{I}} + \kappa_1 \underline{\underline{b}} + \kappa_2 \underline{\underline{b}}^{-1}, \tag{12}$$

where the scalars κ_0, κ_1 and κ_2 are only functions of the temperature and the invariants I_1, I_2 and I_3 of the internal Finger tensor $\underline{\underline{b}}$. In equilibrium state the internal Finger tensor equals $\underline{\underline{b}} = \underline{\underline{I}}$ and the heat conduction tensor reduces to

$$\begin{aligned} \underline{\underline{\kappa}} &= \kappa_{\text{eq}} \underline{\underline{I}}, \\ \kappa_{\text{eq}} &= \kappa_0 + \kappa_1 + \kappa_2, \end{aligned} \tag{13}$$

where κ_{eq} is the heat conduction coefficient that would be measured in the equilibrium state. If $\kappa_2 = 0$ is taken in (12) the heat conduction tensor reduces to the heat conduction tensor derived from microrheological considerations by van den Brule (1989, 1990) for a Maxwell fluid.

If no interaction between different modes is assumed the heat conduction tensor for multi-mode models may be written as

$$\underline{\underline{\kappa}} = \kappa_0 \underline{\underline{I}} + \sum_{k=1}^{K} \left(\kappa_{1,k} \underline{\underline{b}}_k + \kappa_{2,k} \underline{\underline{b}}_k^{-1} \right), \tag{14}$$

where K is the number of modes. In principle $2K + 1$ coefficients for the heat conduction tensor have to be determined as a function of temperature and the invariants of $\underline{\underline{b}}_k$. In equilibrium state the heat conduction tensor reduces to

$$\begin{aligned} \underline{\underline{\kappa}} &= \kappa_{\text{eq}} \underline{\underline{I}}, \\ \kappa_{\text{eq}} &= \kappa_0 + \sum_{k=1}^{K} \kappa_{1,k} + \kappa_{2,k}. \end{aligned} \tag{15}$$

From the restriction that the production of entropy Π_s must be positive for all possible processes, it follows that the contribution of the heat flux must be positive:

$$T\Pi_s = T^{-1} \underline{\underline{\kappa}} : \nabla T \nabla T + \text{o.t.} \geq 0. \tag{16}$$

For a derivation see for example Leonov (1987). To fulfil the inequality it is sufficient to require

$$\begin{aligned} \kappa_0 &\geq 0, \\ \left.\begin{aligned} \kappa_{1,k} &\geq 0, \\ \kappa_{2,k} &\geq 0, \end{aligned}\right\} & \quad k = 1, \ldots, K. \end{aligned}$$

In the derivation use has been made of the positiveness of the tensor $\underline{\underline{b}}_k$ and the assumption that there is no mutual interaction between the $\underline{\underline{b}}_k$.

2.2.3. Dissipation

Calculation of a temperature equation from the energy equation is straightforward for a Newtonian fluid. Substitution of $u = c_p T$ in the energy equation then gives an equation for the temperature:

$$\rho c_p \underline{v} \cdot \nabla T = D_m - \nabla \cdot \underline{\phi}_q, \tag{17}$$

where the mechanical dissipation D_m equals the total amount of work: $\underline{\underline{\tau}} : \underline{\underline{d}}$. Viscoelastic fluids however, are able to store elastic energy. These elastic energy contributions do not contribute to the mechanical dissipation.

To determine the dissipative and the elastic part of the energy, we use the matrix formulation used by Jongschaap (1991). Under isothermal conditions Jongschaap assumes that the mechanical dissipation is equal to the stress work minus the change of the free energy f per unit mass:

$$D_m = \underline{\underline{\tau}} : \underline{\underline{d}} - \rho \sum_{k=1}^{K} \frac{\partial f_k}{\partial \underline{\underline{b}}_k} : \dot{\underline{\underline{b}}}_k, \tag{18}$$

where the free energy consists of K independent contributions of each mode. Note that for fully developed flows this definition for the dissipation results in a dissipation which is equal to the stress work.

Jongschaap introduces a way to calculate the free energy, which results in a positive dissipation. It is based on a macroscopic time reversal: a change of sign of $\underline{\underline{d}}$. Then all variables are split in an odd and an even part with respect to this macroscopic time reversal. Since the dissipation has to be positive, it is even with respect to this time reversal. Furthermore the velocity gradient is odd and the change in free energy is even, because it does not depend directly on $\underline{\underline{L}}$. Then we can calculate the odd part of the stress and the even part of the rate of change of the internal deformation tensor, which determine the free energy and the dissipation.

Performing the calculation for the Giesekus model (7) gives the following expression for the dissipation and the free energy

$$\begin{aligned} D_m &= 2\eta_s \underline{\underline{d}} : \underline{\underline{d}} + \sum_{k=1}^{K} \frac{\eta_k}{2\lambda_k^2} \left((1-\alpha_k)\left(I_{1,k} + \operatorname{tr} \underline{\underline{b}}_k^{-1} - 6\right) + \alpha_k \left(\underline{\underline{b}}_k : \underline{\underline{b}}_k - 2I_{1,k} + 3 \right) \right), \\ \rho f &= \sum_{k=1}^{K} \frac{\eta_k}{\lambda_k} \left(I_{1,k} - \ln I_{3,k} \right). \end{aligned} \tag{19}$$

3. Numerical implementation

In this section we describe the numerical methods we have used to solve the equations described in section 2. The system of equations has been decoupled into 3 parts, which are solved iteratively. Each iteration consists of three steps. In the first step we perform an iteration for the equations of motion. In the second step we perform an iteration for the temperature equation. In the last step we perform an iteration for the stress equation. The details of these three steps of the iteration scheme can be found in subsections 3.1, 3.2 and 3.3. The equations of motion and the temperature equation form the elliptic part of the problem. They have been solved with a standard finite element method. The stress constitutive equation is a hyperbolic equation. It has been solved with a streamline integration method. In the following subsections we give for each equation a short description of its numerical implementation and some problems that arise.

3.1. Equations of Motion

To solve the equations of motion we have used an extended quadratic element: the well known Crouzeix-Raviart element. See figure 1. The velocities in the centre point and the pressure derivatives can be eliminated on element level. To eliminate the pressure p in the centre point the penalty method has been used. Instead of the divergence equation (1)

$$\epsilon_p p + \nabla \cdot \underline{v} = 0, \tag{20}$$

is used. The penalty parameter ϵ_p is a small parameter, such that $\epsilon_p p$ is of the order 10^{-6} to 10^{-8}.

To solve the discretized equations of motion, we use an incremental formulation for the resulting matrix-vector equation:

$$\left(\eta_{\text{it}} S_v + M_v(U^i) + C_v\right) \Delta U^{i+1} = -R_v(U^i, \Upsilon^i, T^i), \tag{21}$$

where the Picard iteration matrix contains contributions of a proper linearisation of the convective terms M_v, the penalty matrix C_v and a viscous matrix $\eta_{\text{it}} S_v$. The iteration viscosity η_{it} only slows down the iteration process. For our calculations in section 5 we used $\eta_{\text{it}} = 2\eta_0$, where η_0 is the zero-shear-rate viscosity at the lowest temperature in the flow. The increment of the velocities ΔU^{i+1} is defined by

$$\Delta U^{i+1} = U^{i+1} - U^i. \tag{22}$$

The residual $R_v(U^i, \Upsilon^i, T^i)$ depends on the discrete velocities U^i, temperatures T^i and stresses Υ^i of a former iteration step:

$$\begin{aligned} R_v(U^i, \Upsilon^i, T^i) &= \left(\eta_s(T^i) + \eta_{\text{co}}\right) S_v U^i + N_v(U^i) + \\ &\quad C_v U^i - F_v + Q(\Upsilon^i - 2\eta_{\text{co}} \bar{D}^i). \end{aligned} \tag{23}$$

N_v contains the contribution of the convective terms, F_v the contributions of the boundary integrals due to the natural boundary conditions and Q the contributions of the divergence of the substresses. From this last term we have to subtract an extra diffusive term based on the nodal point averages of the velocity gradient $\bar{D}^i$. An analogous term based on the non-averaged velocity gradients is added to the contribution of the solvent. This extra viscosity is necessary to avoid almost zero effective viscosities, which may arise due to the use of averaged velocity gradients for streamline integration. Introduction of η_{co} improves the condition of the system but increases the viscosity of 'short wavelength' velocity modes.

For checking convergence two conditions have to be satisfied. The first condition is related to the increment of the velocity and the second to the residual of the momentum equation:

$$\begin{aligned} \frac{\max_j |\Delta U_j^{i+1}|}{\max_j |U_j^{i+1}|} &\leq \epsilon_{v,\text{inc}}, \\ \frac{||R_v(U^i, T^i)||_f}{||R_v(U^i, T^i)||_t} &\leq \epsilon_{v,\text{res}}, \end{aligned} \tag{24}$$

where the maximum norm has been taken over all nodal points j. The norm $||\cdot||_f$ denotes the Euclidean norm over the free degrees of freedom, without the essential

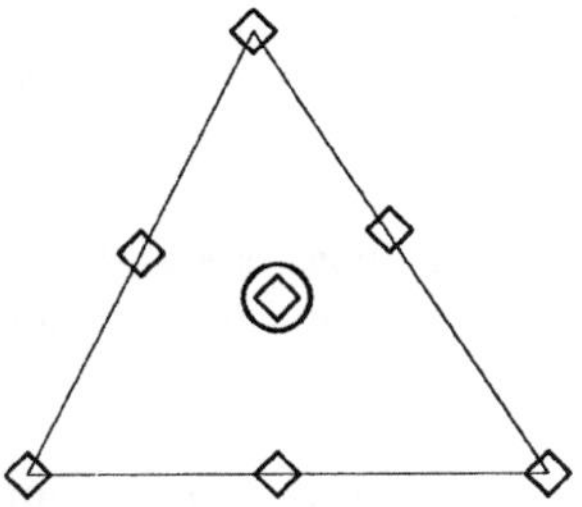

Fig. 1. Crouzeix-Raviart element with seven velocity points (diamonds) and one point with the pressure and its derivatives (circle).

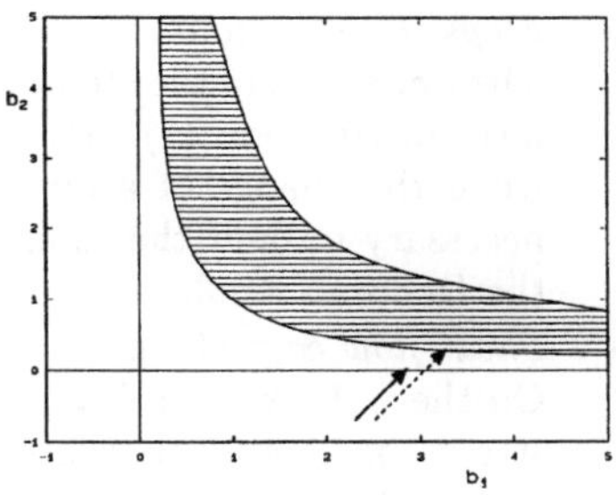

Fig. 2. Correction of the internal deformation tensor in 2D. b_1 and b_2 are the principal values of $\underline{\underline{b}}$. The solution of the model lies in the dashed area, which is bounded by two lines with constant I_3. The straight arrow is the projection on the first quadrant used by Hulsen & van der Zanden (1991). The dashed arrow is a projection on the lowest possible I_3, which is necessary to calculate the dissipation.

boundary conditions. The norm $||\cdot||_t$ denotes the Euclidean norm over the total degrees of freedom, including the essential boundary conditions. For our calculations in section 5 we have taken $\epsilon_{v,\text{inc}} = \epsilon_{v,\text{res}} = 10^{-3}$.

3.2. Stress Equations

The Giesekus model can be written as a set of ordinary differential equations for the internal deformation tensor, by applying the method of characteristics

$$\frac{d\underline{\underline{b}}_k}{ds}\frac{ds}{dt} = -\frac{1}{\lambda_k(T)}\left(\underline{\underline{L}}\cdot\underline{\underline{b}}_k + \underline{\underline{b}}_k\cdot\underline{\underline{L}}^T + \underline{\underline{I}} + \alpha_k\left(\underline{\underline{b}}_k - \underline{\underline{I}}\right)\cdot\left(\underline{\underline{b}}_k - \underline{\underline{I}}\right)\right), \qquad (25)$$

where s is a streamline parameter. Integration is performed with a fourth order Runge-Kutta scheme. Below we summarize some important aspects of the streamline integration method and the difficulties that may arise. More details about the isothermal streamline integration can be found in Hulsen & van der Zanden (1991).

- *Computation of the streamlines.*
  From the quadratic velocities of the FEM-part we calculate a quadratic stream function. With this stream function we calculate a piecewise quadratic streamline.
- *Initial conditions.*
  For the integration of (25) we have to specify an initial condition for the internal deformation tensor. There are 3 possibilities. Firstly the streamline may cross an inflow boundary. Secondly the streamline may cross an element where it has already been computed. Finally, to reduce the computing time, the integration is stopped if the travel time is longer than $3\lambda_k(T)$, the relaxation time at the starting point. The initial values are then found by interpolation from the nodal point values of the previous iteration.

- *Stepsize limitations.*
  The limitation of the stepsize of streamline integration is based on the linearized form of equation (7). Furthermore it is possible to limit the stepsize further when the change of a quantity during a step is too large. For example it is necessary to limit the change in velocity gradient at start-up for high values of the Deborah number.
- *Shear flow correction.*
  On the wall numerical approximation errors may cause large false elongational stresses for high Deborah numbers. This can be avoided by imposing an exact simple shear flow on the wall.
- *Correction of the internal deformation tensors.*
  Due to numerical approximation errors the internal deformation tensor can become indefinite in regions where large gradients are present, for example near sharp corners. Due to the long residence time near the wall, negative determinants may blow up the quadratic term in (7). To avoid these nonlinear instabilities we correct the internal deformation tensor with an isotropic term. See figure 2. For the stress equation it is sufficient to put $\underline{\underline{b}}$ back on one of the axes, so it becomes semi-positive definite. However for the calculation of the dissipation this is not sufficient, because we need the inverse of the $\underline{\underline{b}}$ tensor in equation (19). For some other models it is possible to find a sharper lower bound for the determinant of $\underline{\underline{b}}$. For the Leonov model, for example, the determinant always equals $I_3 = 1$ and for the Phan-Thien/Tanner model $I_3 \geq 1$. This can be used to make a projection on the curve $I_3 = 1$ instead of one of the principal axes. For the 2D Giesekus model it is also possible to find a lower bound which depends on the parameter α in the Giesekus model (see Hulsen (1988)). However for the 3D axisymmetrical Giesekus model we were not able to find such a lower bound and had to proceed differently. If the determinant of the internal deformation tensor is smaller than 10^{-6} we took the stress work of the specific mode instead of the dissipation. In our examples in section 5 we had to take the stress work for points near the sharp corner of the first two modes, the ones with the largest elasticity.

3.3. Temperature Equation

To solve the temperature equation we have used a quadratic element. For the interpolation of quantities to the Gauss integration points we use quadratic interpolation. The quadratic interpolation may destroy the positive definiteness of the internal deformation tensor. This can give rise to large numerical errors in terms where the inverse of $\underline{\underline{b}}$ has to be calculated. Therefore all quantities that depend on $\underline{\underline{b}}^{-1}$ must first be calculated in the nodal points and can only then be interpolated to the integration points.

For convection dominated problems the standard Galerkin method can give rise to unphysical solutions, with a globally oscillating character. For our type of problems the temperature equation is often convection dominated due to the small value of the diffusivity of a viscoelastic material. The value for our examples in section 5 is $\frac{\kappa}{\rho c_p} = \mathcal{O}(10^{-7})$. To avoid the unphysical wiggles the value of the mesh Péclet number must fulfil in all elements the severe condition $Pe_m = \frac{\rho c_p v \Delta x}{\kappa_v} \leq 2$, where v,

Δx and κ_v are the velocity, the size of the element and the heat conduction in the streamwise direction. To fulfil this restriction the size of the elements of the mesh must be extremely small. This limits of course the practical use of the Galerkin method. To avoid extreme mesh refinement for high values of the mesh Péclet number an upwind technique can be used. The wiggles can be suppressed with help of such methods, but they can cause inaccurate solutions through the introduction of false (extra) diffusion, especially for coarse grids. The false diffusion reduces when the element size becomes smaller.

For the consistent upwind methods all terms are multiplied with an upwind test function ψ_u and added to the weak formulation of the temperature equation.

$$\int_\Omega (\rho c_p \underline{v} \cdot \nabla T - D_m - \nabla \cdot (\underline{\underline{\kappa}} \cdot \nabla T))(\psi + \psi_u)\, d\Omega = 0, \qquad \forall \psi \in \Psi^t, \quad \forall \psi_u \in \Psi^t_u. \tag{26}$$

These upwind testfunctions consist of an upwind direction that determines the direction of the upwinding and an upwind function that determines the amount of upwinding.

The Streamline upwind Petrov-Galerkin method (SUPG) has been described by Hughes & Brooks (1982). They take the streamlines as the upwind direction,

$$\psi_u = \tau \underline{v} \cdot \nabla \psi, \tag{27}$$

so this upwind method shows no false crosswind diffusion. With this choice of ψ_u sufficiently smooth exact solutions can be approximated very well. The presence of sharp layers can create local oscillations, in contrary to the standard Galerkin method which creates globally-propagating oscillations. The local wiggles can be suppressed by adding some diffusion in the direction of the temperature gradient as described by Mizukami & Hughes (1985). However for anisotropic heat conduction this method does not work, because the direction of the heat flux does not equal the direction of the temperature gradient.

The upwind function τ indicates the amount of upwinding. It is based on an element Péclet number β_u

$$\begin{aligned} \beta_u &= \frac{||\underline{v}|| h \rho c_p}{2\kappa_v}, \\ \kappa_v &= \frac{\underline{v} \cdot \underline{\underline{\kappa}} \cdot \underline{v}}{||\underline{v}||^2}, \end{aligned} \tag{28}$$

where h is the maximum distance in the element in the direction of $\underline{v}$ and κ_v is a measure for the heat conduction in the streamwise direction. If $\beta_u < 1$ the upwind function can be chosen small, because the standard Galerkin approach still gives accurate solutions. For $\beta_u > 1$ however, upwinding has to be applied to avoid unphysical wiggles. Several choices for τ are used in the literature. We have used the 'optimal' upwind function for the examples in section 5.

$$\begin{aligned} \tau &= \frac{h \xi_u}{2||\underline{v}||}, \\ \xi_u &= \coth \beta_u - \frac{1}{\beta_u}, \end{aligned} \tag{29}$$

where ξ_u is a non-dimensional numerical diffusivity. The choice of τ seems not of great importance. In our test examples we could not find significant differences between different choices of τ.

To solve the discretized temperature equation we also use an incremental formulation. All of the matrices and vectors may contain contributions of the upwind scheme:

$$\left(\kappa_{\rm it} S_T(\Upsilon^i) + M_T(U^i, T^i)\right) \Delta T^{i+1} = -R_T(U^i, \Upsilon^i, T^i). \tag{30}$$

The Picard iteration matrix contains contributions of the convective terms M_T and a diffusive matrix $\kappa_{\rm it} S_T$. The iteration diffusivity $\kappa_{\rm it}$ slows down the iteration process. This avoids convergence problems. A constant and isotropic iteration diffusivity gives good results for our examples in section 5. The increment ΔT^{i+1} is defined by

$$\Delta T^{i+1} = T^{i+1} - T^i, \tag{31}$$

and the residual $R_T(U^i, \Upsilon^i, T^i)$ equals

$$R_T(U^i, \Upsilon^i, T^i) = S_T(\Upsilon^i) T^i + N_T(U^i, T^i) - F_T(U^i, \Upsilon^i, T^i). \tag{32}$$

where the diffusive matrix $S_T(\Upsilon^i)$ may depend on the stresses due to the anisotropic heat conduction. N_T contains the contributions of the convective terms. The dissipation and the boundary integrals of the natural boundary conditions contribute to F_T.

For checking convergence two conditions analogous to the momentum equations have to be satisfied. The first condition is related to the increment of the temperature and the second to the residual of the temperature equation:

$$\begin{aligned} \frac{\max_j |\Delta T_j^{i+1}|}{\max_j |T_j^{i+1}|} &\leq \epsilon_{T,\rm inc}, \\ \frac{||R_T(U^i, T^i)||_f}{||R_T(U^i, T^i)||_t} &\leq \epsilon_{T,\rm res}, \end{aligned} \tag{33}$$

where the maximum norm has again been taken over all nodal points j. The norms $||\cdot||_f$ and $||\cdot||_t$ again denote the Euclidean norm over the free and total degrees of freedom. For our calculations presented in section 5 we have taken $\epsilon_{T,\rm inc} = \epsilon_{T,\rm res} = 10^{-3}$.

3.4. Boundary conditions

This last part of the numerical implementation deals with the boundary conditions. Especially the boundary conditions at the inflow and outflow:

– *Inflow boundary.*
  At an inflow boundary we have to specify the velocity, the temperature and all substresses. We assume that these quantities are fully developed. For the calculation we have a simple 1D program. We decouple the temperature equation and the equations of motion. This results in an algebraic equation for the shear rate and an ordinary differential equation for the stresses. The algebraic equation for the shear rate is solved with a simple Picard or secant method. To obtain a convergent scheme we had to add an iteration diffusivity in the

integration of the temperature equation. This procedure is analogous to the finite element temperature equation.

- *Outflow boundary.*
  In principle it is possible to impose fully developed boundary conditions at the outflow, as we do for the isothermal calculations. For nonisothermal flows with high Péclet numbers, however, this would require very long exit lengths. This would require long computation times and a large memory capacity of your computer. To avoid this, we impose for the outflow natural boundary conditions that are based on the fully developed flow. The procedure is as follows. For the temperature we suppose a zero heat flux in the normal direction. For the momentum equations we prescribe the normal stress and a zero tangential velocity. The only problem is to find a good approximation of the normal stress at the outflow. We have done that by using the temperature resulting from the natural temperature boundary conditions for calculation of the viscosities and relaxation times. With these calculated viscosities and relaxation times we calculate a fully developed pressure and a fully developed extra stresses, which form the normal stress at the outflow.

4. Problem Description

By Hulsen & van der Zanden (1991) an eight-mode Giesekus model has been used to calculate the steady, isothermal flow of a low density polyethylene (LDPE) melt through a 4 to 1 contraction. For the nonisothermal calculations we have also used this model described by Bird et al. (1987). The viscoelastic material parameters for this model can be found in table I. The thermal properties in table II are also from Bird et al. (1987). So far we know there are no measurements of the anisotropical heat conduction parameters of LDPE. For our calculations we therefore assumed some parameters, which are given in table III. This means that 80% of the heat conduction takes place through the polymer chains. This value is not unusual for rubbers, see van den Brule (1989). In figure 3 and 4 we have plotted the shear viscosity and the elongational viscosity for this model for the temperature range we use in our calculations in section 5.

The temperature boundary conditions we have used for our axisymmetrical examples in section 5 have been plotted in figures 5 and 6. In the first example the wall near the vortex has been heated. In the second example the wall has been cooled from the contraction to the outflow. On the fixed wall we have assumed that the no-slip boundary condition holds for the equations of motion. For the inflow and outflow we have prescribed fully developed profiles as described in subsection 3.4.

For our calculations we varied two dimensionless numbers, the Deborah number and the Péclet number. Because of the problems with the calculation of the dissipation for the 3D Giesekus model we didn't look more closely at the influence of the dissipation. In our examples the temperature rise due to dissipation is small, less than $0.5K$. Both the dimensionless numbers are calculated at the outflow. The Deborah number, which denotes the ratio of the characteristic time scale of the fluid

TABLE I
Viscoelastic properties of LDPE, from Bird et al. (1987)

k	$\eta_k(\mathrm{Pa\cdot s})$	$\lambda_k(\mathrm{s})$	α_k
1	$1.00\cdot 10^3$	10^3	0.03
2	$1.80\cdot 10^4$	10^2	0.05
3	$1.89\cdot 10^4$	10^1	0.2
4	$9.80\cdot 10^3$e	10^0	0.5
5	$2.67\cdot 10^3$	10^{-1}	0.4
6	$5.86\cdot 10^2$	10^{-2}	0.3
7	$9.48\cdot 10^1$	10^{-3}	0.2
8	$1.29\cdot 10^1$	10^{-4}	0.1

TABLE II
Thermal properties of LDPE, from Bird et al. (1987)

ρ	$7.82\cdot 10^2\mathrm{kg/m^3}$
c_p	$2.57\cdot 10^3\mathrm{J/kgK}$
κ	$2.41\cdot 10^{-1}\mathrm{W/mK}$
T_{ref}	423K
c_1	$4.50\cdot 10^3\mathrm{K}$

TABLE III
Heat conduction constants in (W/mK) for all modes k

κ_0	$4.82\cdot 10^{-2}$
$\kappa_{1,k}$	$2.41\cdot 10^{-2}$
$\kappa_{2,k}$	0

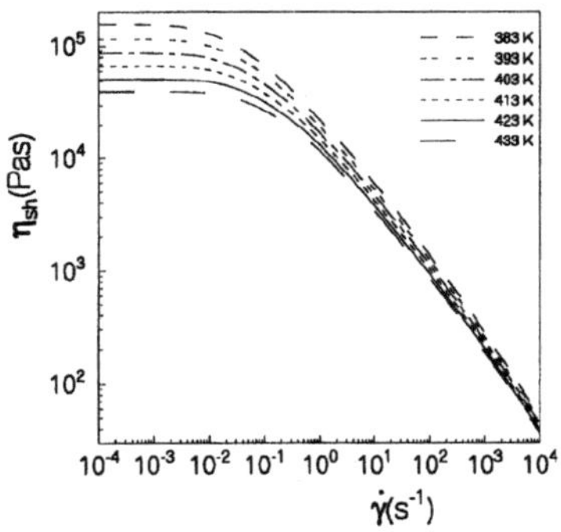

Fig. 3. Shear viscosity for the LDPE melt versus shear rate at different temperatures.

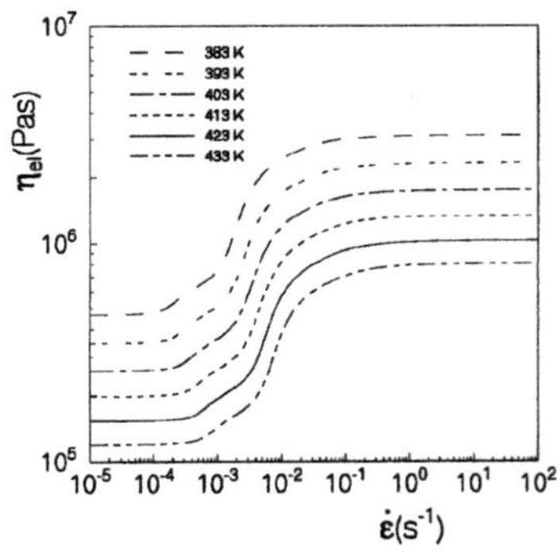

Fig. 4. Elongational viscosity for the LDPE melt versus shear rate at different temperatures.

and a characteristic time scale of the flow, is defined by

$$De = \frac{\lambda_{0,\mathrm{ref}}\langle v\rangle}{D_d},$$

with $\langle v\rangle$ the average velocity and D_d the diameter at the outflow. The mean relaxation time at reference temperature $\lambda_{0,\mathrm{ref}}$ is defined as

$$\lambda_{0,\mathrm{ref}} = \sum_{k=1}^{K} \frac{\lambda_{k,\mathrm{ref}}\eta_{k,\mathrm{ref}}}{\eta_{0,\mathrm{ref}}},$$

with $\eta_{0,\mathrm{ref}} = \eta_{s,\mathrm{ref}} + \sum_{k=1}^{K}\eta_{k,\mathrm{ref}}$ the zero-shear-rate viscosity at reference temper-

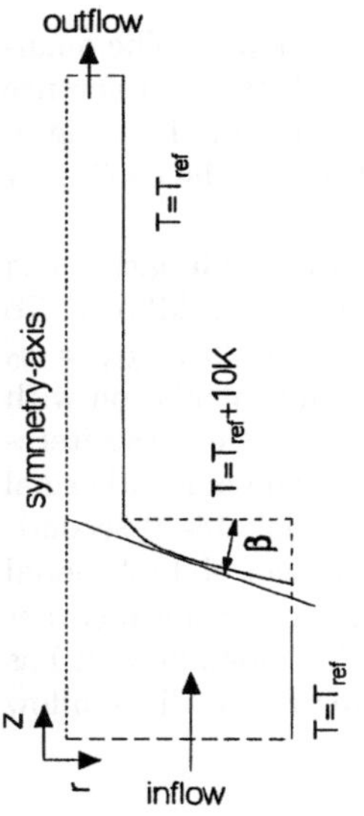

Fig. 5. Boundary conditions for problem 1 and definition of the opening angle β. The wall near the vortex (dashed line) has been heated. The other part of the wall has been kept at reference temperature.

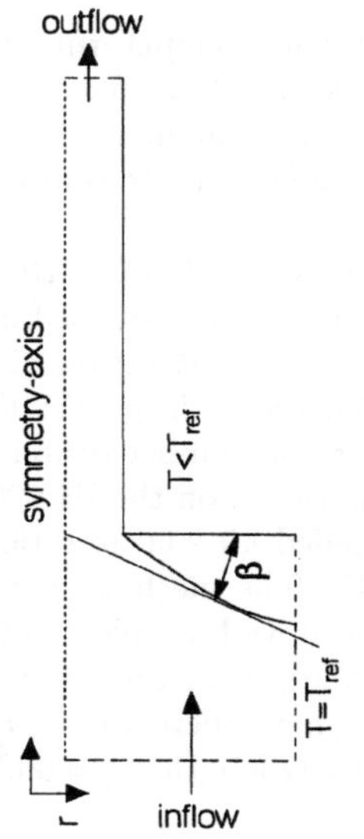

Fig. 6. Boundary conditions for problem 2. The wall at the inflow (dashed line) has been kept at reference temperature. The other part of the wall (straight line) has been cooled.

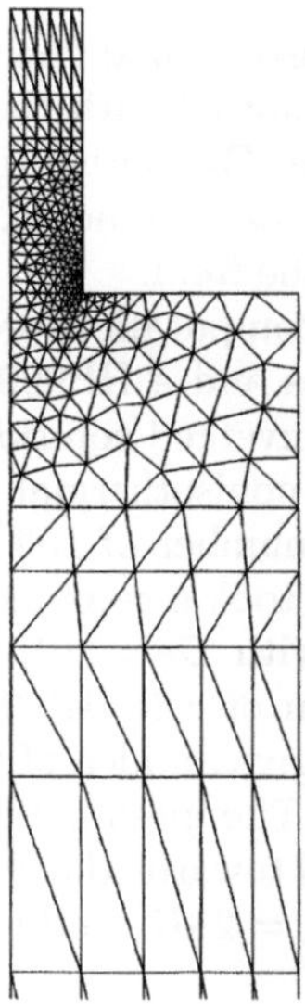

Fig. 7. A part of the finite element mesh near the sharp corner.

ature. The Péclet number, denoting the ratio of the convective transport and the transport by conduction, is based on the equilibrium thermal conductivity coefficient:

$$Pe = \frac{\rho c_p \langle v \rangle D_d}{2\kappa_{\mathrm{eq}}}.$$

We have examined three characteristic quantities for the flow through a contraction. Firstly the opening angle β, defined in figure 5. The second quantity is the vortex intensity, the ratio of the amount of fluid flowing in the vortex and in the main flow:

$$I_\psi = \frac{\psi_{\mathrm{sep}} - \psi_{\mathrm{cen}}}{\psi_{\mathrm{axis}} - \psi_{\mathrm{sep}}},$$

where ψ is the stream function value at the separating streamline ψ_{sep}, the centre of the vortex ψ_{cen} and the symmetry axis ψ_{axis}. Finally we examine an entrance correction, defined by

$$n_c^{\mathrm{iso}} = \frac{-\sigma_n^{\mathrm{en}} + \sigma_n^{\mathrm{ex}} - \left(\Delta p_{\mathrm{en}}^{\mathrm{dev}} + \Delta p_{\mathrm{ex}}^{\mathrm{dev}}\right)^{\mathrm{iso}}}{2\tau_w^{\mathrm{iso}}}.$$

σ_n^{en} and σ_n^{ex} are the total normal stresses at the wall of the entry and exit. $\Delta p_{\mathrm{en}}^{\mathrm{dev}}$ and $\Delta p_{\mathrm{ex}}^{\mathrm{dev}}$ are the pressure differences in the entrance and exit region that would exist if the isothermal flow at reference temperature was fully developed. The quantity is normalized with the wall shear stress from the isothermal flow at reference temperature. For isothermal flows this quantity gives the extra pressure loss due to the contraction. For nonisothermal flows there is also an effect due to the cooling or heating of the fluid.

For our computations we have used a HP9000-720 computer with a Linpack speed of 14 Mflops and a HP9000-735 computer with a Linpack speed of 40 Mflops. To obtain a converged solution we used an iteration viscosity of $\eta_{\text{it}} = 2\eta_0$, $\eta_{\text{co}} = \eta_0$ and for the nonisothermal problems $\kappa_{\text{it}} = 5\kappa_{\text{eq}}$. For the isothermal calculation with a Deborah number of 100 the largest number of iterations were needed, 189 iterations which took 6 seconds per iteration on the HP9000-735. For the nonisothermal problems, with $De = 10$, we needed 40 - 60 iterations, which took 15-20 seconds per iteration on the HP9000-735. The mesh we have used consists of 1452 nodal points. To give an idea of the mesh we have plotted the part near the sharp corner in figure 7. The mesh is rather coarse near the inflow and outflow boundary. It has been refined towards the sharp corner, where large gradients are present. The inflow length is $L_u = 20D_d$ and the outflow length $L_d = 60D_d$.

5. Results

In this section we will examine the influence of the Deborah number, the Péclet number and the cooling temperature on the flow. Figure 8 shows the influence of the Deborah number for isothermal flows. The opening angle β becomes larger when the Deborah number increases. The idea is that the growth of the vortex is a mechanism to fulfil the balance of momentum in the z-direction:

$$\frac{1}{r}\frac{\partial r\tau_{rz}}{\partial r} + \frac{\partial \tau_{zz}}{\partial z} \simeq 0, \tag{34}$$

where we have neglected the pressure gradient. The build-up of the dominant term in this equation, τ_{zz}, before the contraction can be more gradually with a larger vortex. For low Deborah numbers the vortex intensity increases with increasing De. For high Deborah numbers the vortex intensity slowly decreases with increasing De. Some numerical values can be found in table IV.

For all the nonisothermal examples with boundary conditions defined in figure 5 and 6 the opening angle didn't change significantly. The influence of the temperature changes was too small or too locally. The differences in the vortex intensities, however, are relatively large.

For problem 1 the heating has two opposite effects. On the one hand the vortex intensity is decreased. An increase in the temperature gives rise to a fluid that is less elastic. A temperature rise of $10K$ of the whole fluid with $De = 10$ corresponds to an isothermal flow with $De = 7.8$. In this region the vortex intensity increases with increasing elasticity, see table IV. On the other hand the vortex intensity will be increased due to the temperature differences in the flow. Due to the higher temperatures in the vortex region, the viscosity is locally smaller than in the main flow. This results in larger velocities in the vortex and consequently larger vortex intensities. In table V we see that for the isotropic heat conduction the first effect is dominant for the lowest Péclet number and the second effect is dominant for the

TABLE IV
Vortex intensity and opening angle for different Deborah numbers

De	$I_\psi(\%)$	$\beta(\deg)$
1	1.8	25
7.8	10.8	47
10	11.3	47
100	10.5	58

TABLE V
Vortex intensity for different Péclet numbers for problem 1. The first column is without anisotropy, the second with the anisotropy parameters from table III. The Deborah number equals $De = 10$.

Pe	$I_\psi(\%)$	$I_\psi^a(\%)$
10	11.1	12.1
100	12.4	12.4
1000	12.6	12.6

others. The reason for this can be seen in figure 9. For the lowest Pe the temperature difference between the vortex and the main flow is relatively small, in contrary to the other Péclet numbers. In table V and figure 10 the influence of the anisotropy on the vortex intensity and the temperature field can be found. For the smallest Péclet number the influence of the anisotropy is relatively large. The anisotropy of the heat conduction influences the temperature field near the contraction drastically. Due to the large elongational stresses the polymer chains will be oriented in the direction of the flow. This results in an increase of the heat flux in the direction of the flow and a decrease in the direction perpendicular to it. This results in a larger temperature difference between the vortex region and the main flow, and consequently a larger vortex intensity. For $Pe = 100$ and $Pe = 1000$ the flow is too convection dominant to give large differences between the figures 9 and 10. There are only small differences near the sharp corner, where the stresses are relatively large and the velocity is relatively small.

In the second example we will examine the influence of the cooling temperature. Again the influence on the opening angle is relatively small. The influence on the vortex intensity and the entrance correction has been summarized in table VI. For these calculations $De = 10$, $Pe = 1000$ and anisotropy is taken into account. Due to the cooling of the wall the temperature in the vortex and outflow region becomes lower. In the vortex region, however, the cooling is more effective. Firstly because the velocity is smaller and thus the local Péclet number is smaller in the vortex region. Secondly because the stresses in the outflow region are larger, which results in a stronger orientation of the polymer chains. This gives a reduction in the heat conduction in the r-direction. This results in a decreasing vortex intensity for a decreasing wall temperature. Additionally we calculated the entrance corrections for the different cooling temperatures. Due to the low thermal diffusion perpendicular to the flow direction compared to the convective transport parallel to the flow the cold temperature boundary layer develops very slowly along the wall. Consequently the temperature profile is far from fully developed at the outlet and near the symmetry axis there is no decrease of the temperature at all. This can be seen in figure 11, where the temperature isolines have been plotted for 1/6 of the exit length and 3/10

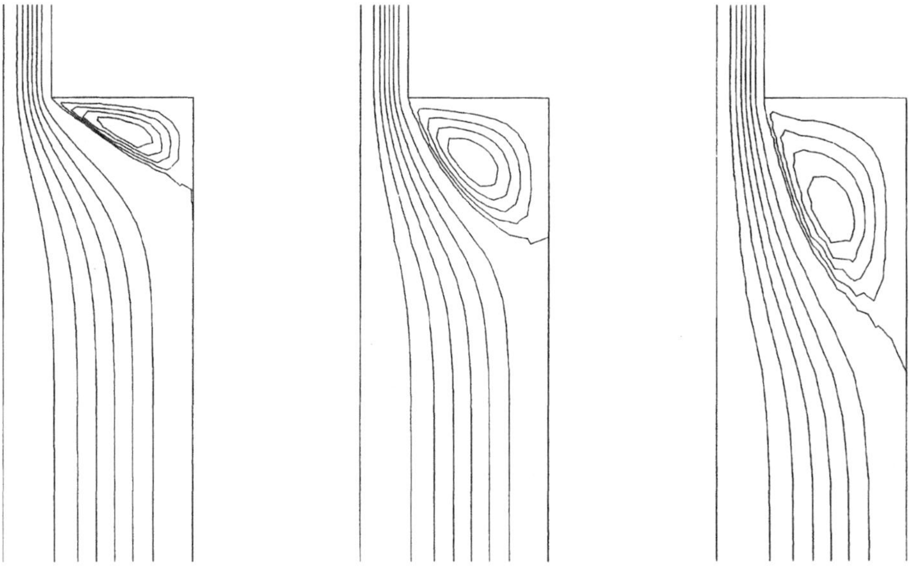

Fig. 8. Streamlines for isothermal flows with increasing Deborah number. $De = 1$, $De = 10$ and $De = 100$.

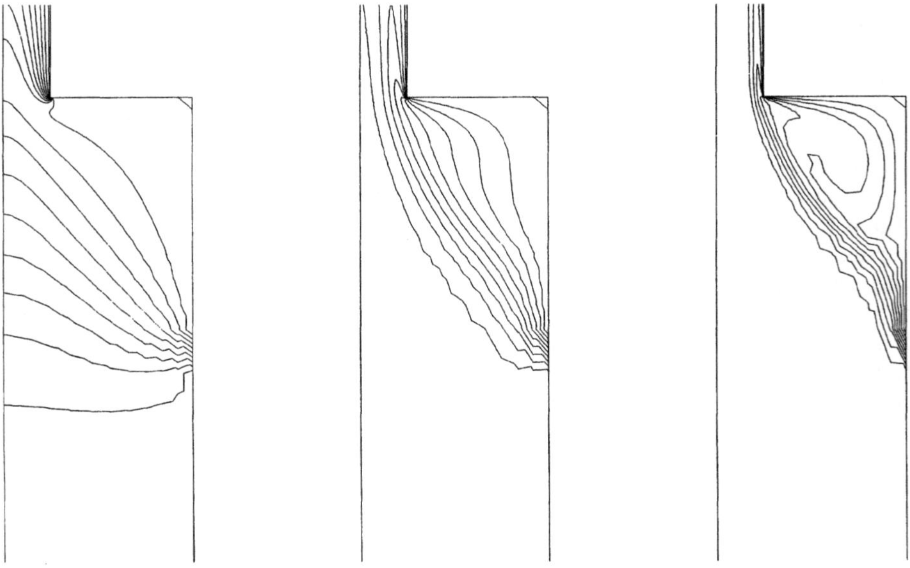

Fig. 9. Temperature isolines for increasing Péclet numbers for problem 1. $Pe = 10$, $Pe = 100$ and $Pe = 1000$, without anisotropic heat conduction. Each isoline differs about $0.9K$ from the next one.

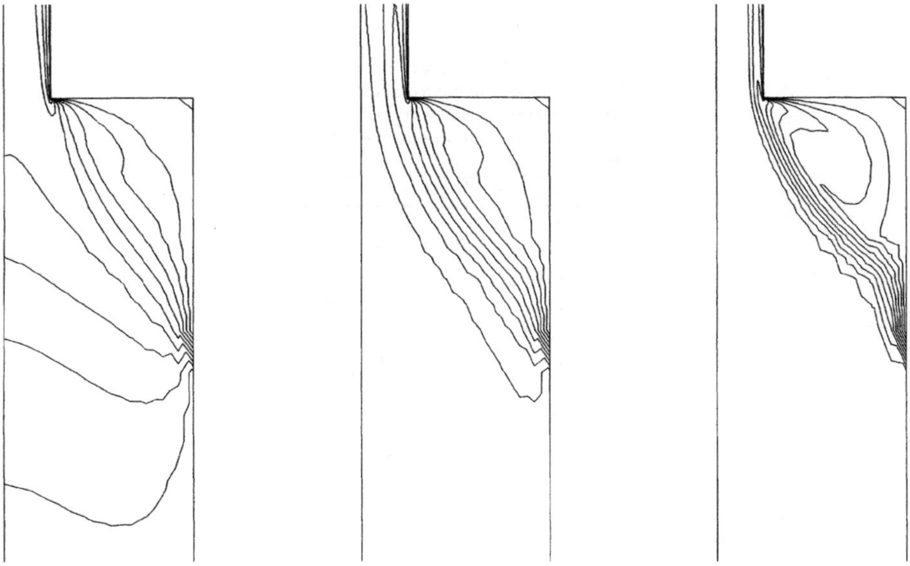

Fig. 10. Temperature isolines for increasing Péclet numbers for problem 1. $Pe = 10$, $Pe = 100$ and $Pe = 1000$, with anisotropic heat conduction. Each isoline differs about $0.9K$ from the next one.

of the entry length. The isoline nearest to the axis of symmetry is only $1K$ below the reference temperature. Figure 11 also shows that a relatively large part of the cold layer in the exit is due to the convection from the entry region. In this region the local Péclet number is smaller, so the cold front can penetrate more easily in the main flow there. Although the cold temperature layer is relatively small in radial direction the influence on the magnitude of the entrance correction is large compared to the isothermal flow. For the lowest wall temperature the entrance correction is more than 10 times larger than for the isothermal flow. Note that the magnitude of n_c^{iso} depends strongly on the exit length, the longer the exit length the larger n_c^{iso}. Of course it would be better to have a quantity that does not depend on the exit length. This is however very difficult, if not impossible, because the cold layer is determined by both convection from the vortex region and diffusion from the cold wall in the exit.

6. Concluding remarks

We have extended the numerical implementation of Hulsen & van der Zanden (1991) for steady and isothermal flow of viscoelastic fluids with some nonisothermal effects. The viscosities and relaxation times have been made temperature dependent by means of a shift factor. In the energy equation two effects that are special for viscoelastic fluids have been taken into account. The first is the anisotropy of the heat conduction due to molecular orientation of the polymer chains. The second is storage

TABLE VI
Shift factor at the cooled wall, vortex intensity and entrance correction for different cooling temperatures at the wall.

T_{wall}(K)	a_T^{wall}	I_ψ(%)	n_c^{iso}
423	1.00	11.3	1.5
413	1.29	10.1	5.6
403	1.70	8.8	10.3
393	2.25	7.4	15.4
383	3.04	6.1	20.9

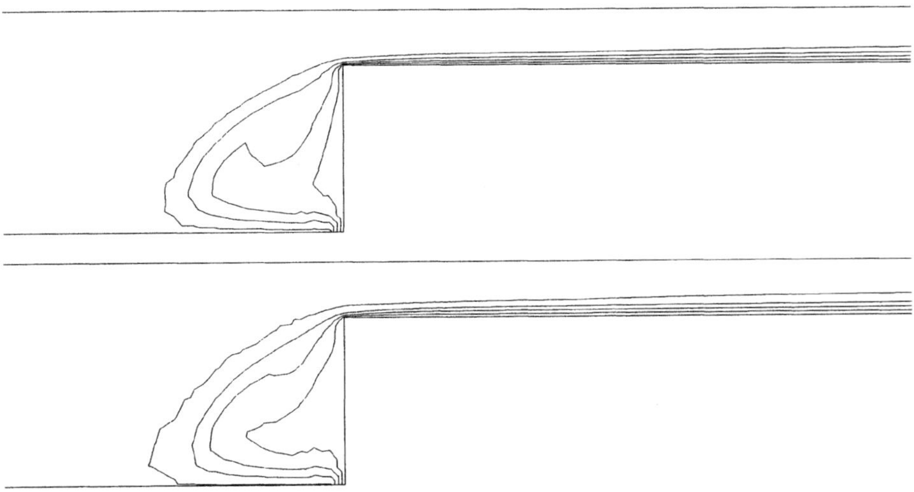

Fig. 11. Temperature isolines for different wall cooling temperatures (30% of the entry length and 16% of the exit length). For $T_{\text{wall}} = 413K$ the isolines are between $T = 413K$ and $T = 422K$. For $T_{\text{wall}} = 383K$ the isolines are between $T = 383K$ and $T = 422K$.

of elastic energy, through which the dissipation does not equal the stress work. For the axisymmetrical Giesekus model we used, we have still some problems for the calculation of this quantity near sharp corners. This is caused by the indefiniteness of the internal deformation tensor.

In our examples we examined the influence of the Péclet number, the anisotropical heat conduction and the wall cooling on the flow. The influence on the opening angle and the main flow velocity is small. The vortex intensity and the entrance correction, however, show large differences. The influence of the anisotropical heat conduction is relatively small for high Péclet numbers.

Acknowledgements

We want to acknowledge J.P.P.M. van der Zanden for his advices and help with the extension of the numerical code to nonisothermal flows and G.D.C. Kuiken for his comments on this work.

References

Bird, R.B., Armstrong, R.C. & Hassager, O.: 1987, 'Dynamics of polymeric liquids', Vol. 1 2nd edn., Wiley, New York

Brule, B.H.A.A. van den: 1989, 'A network theory for the thermal conductivity of an amorphous polymeric material', *Rheol. Acta* Vol. no. **28** pp. 257-266

Brule, B.H.A.A. van den: 1990, 'The non-isothermal elastic dumbbell: a model for the thermal conductivity of a polymer solution', *Rheol. Acta* Vol. no. **29** pp. 416-422

Crochet, M.J.: 1989, 'Numerical simulation of viscoelastic flow: a review', *Rubber Chem. Technol.* Vol. no. **62** pp. 426–455

Giesekus, H.: 1982, 'A simple constitutive equation for polymer fluids on the concept of deformation-dependent tensorial mobility', *J. Non-Newtonian Fluid Mech.* Vol. no. **11** pp. 69–109

Hands, D.: 1980, 'The effect of biaxial orientation on the thermal coductivity of vulcanized and unvulcanized rubber', *Rubber Chem. Technol.* Vol. no. **53** pp. 80–87

Hughes, T.J.R. & Brooks, A.: 1982, 'A theoretical framework for Petrov- Galerkin methods with discontinuous weighting functions: application to the streamline-upwind procedure' *Finite elements in fluids* (ed. Gallagher, R.H., Norrir, D.H., Oden, J.T. & Zienkiewicz, O.C.), Vol. no. 4, Wiley, New York, pp. 47–65

Hulsen, M.A.: 1988, 'Some properties and analytical expressions for plane flow of Leonov and Giesekus models', *J. Non-Newtonian Fluid Mech.* Vol. no. **30** pp. 85–92

Hulsen, M.A.: 1990, 'A sufficient condition for a positive definite configuration tensor in differential models', *J. Non-Newtonian Fluid Mech.* Vol. no. **38** pp. 93–100

Hulsen, M.A. & van der Zanden, J.P.P.M.: 1991, 'Numerical simulation of contraction flows using a multi-mode Giesekus model', *J. Non-Newtonian Fluid Mech.* Vol. no. **38** pp. 183–221

Jongschaap, R.J.J.: 1991, 'Towards a unified formulation of microrheological models', *Lecture Notes in Physics* Vol. no. **381**, pp. 215–247

Keunings, R.: 1989, 'Simulation of viscoelastic fluid flow', *Fundamentals of computer modeling for polymer processing* (ed. Tucker, C.L. III), Carl Hanser Verlag, pp. 403–469

Leonov, A.I.: 1987, 'On a class of constitutive equations for viscoelastic liquids', *J. Non-Newtonian Fluid Mech.* Vol. no. **25** pp. 1–59

Mizukami, A. & Hughes, T.J.R.: 1985, 'A Petrov-Galerkin finite element method for convection dominated flows: an accurate upwinding technique for satisfying the maximum principle', *Comp. Meth. Appl. Mech. Eng.* Vol. no. **50** pp. 181–193

Washo, B.D. & Hansen, D.: 1969, 'Heat conduction in linear amorphous high polymers: orientation anisotropy', *J. Appl. Phys.* Vol. no. **7** pp. 2423–2427

OSCILLATIONS IN PISTON-DRIVEN SHEAR FLOW OF A NON-NEWTONIAN FLUID

DAVID S. MALKUS
University of Wisconsin

JOHN A. NOHEL
ETH-Zürich and University of Wisconsin

and

BRADLEY J. PLOHR
State University of New York at Stony Brook

Abstract. In recent experiments on piston-driven shear flow of a highly elastic and very viscous non-Newtonian fluid, Lim and Schowalter observed nearly periodic oscillations in the particle velocity at the channel wall for particular values of the constant volumetric flow rate. Such periodicity has been characterized as a "stick/slip" phenomenon caused by the failure of the fluid to adhere to the wall. We suggest an alternative explanation for these oscillations using a dynamic mathematical model based on the Johnson-Segalman-Oldroyd constitutive relation, the key feature of which is a non-monotonic relationship between steady shear stress and strain rate. The resulting three-dimensional equations of motion and stress are reduced to one space dimension, consistent with experimental results. In the inertialess approximation, the equations governing the flow can be viewed as a continuous family of quadratic ordinary differential equations coupled by the non-local constraint that fixes the volumetric flow rate. Varying the flow rate, the numerical simulation of solutions to this system exhibits transitions to and from a regime with persistent oscillations that compare favorably with the Lim-Schowalter observations. When the time-asymptotic behavior is cyclic, large shear strain rates are observed in a thin but macroscopic layer near the wall. The transitions are explained using spectral analysis of the linear (infinite dimensional) operator resulting from linearization of the quadratic system about a discontinuous steady state with a jump discontinuity in the stress components. The persistent oscillations arise as a Hopf bifurcation to periodic orbits as the volumetric flow rate is increased beyond a critical value.

Key words: Shear flow, piston-driven, non-Newtonian, non-monotone, oscillations, spurt, feedback control, quadratic systems, Hopf bifurcation, periodic orbits.

1980 *Mathematics Subject Classification* (1985 *Revision*). 34C35, 34K35, 65N30, 76A05, 76A10

This work was supported in part by: the National Science Foundation under Grant DMS-9216153; the National Science Foundation under Grant DMS-9201581; the National Science Foundation under Grant INT-9216357; the U. S. Army Research Office under Grant DAAL03-92-G-0185; the U. S. Army Research Office under Grant DAAL03-91-C-0027 to the Mathematical Sciences Institute of Cornell University, through subcontract to Stony Brook; and the Applied Mathematics Subprogram of the U. S. Department of Energy under Grant DE-FG02-90ER25084.

1. Introduction

In their recent experiments on piston-driven flow at a fixed volumetric flow rate of a highly elastic and very viscous non-Newtonian fluid, F. Lim and W. Schowalter [4, 5] observed persistent, nearly periodic oscillations in the particle velocity near the channel wall. They characterized the periodicity as a stick/slip phenomenon; oscillations are thought to reflect periodic detachment and reattachment of the fluid,

J.F. Dijksman and G.D.C. Kuiken (eds.),
IUTAM Symposium on Numerical Simulation of Non-isothermal Flow of Viscoelastic Liquids, 57-71.

while the regular temporal variation of normal stresses leads to spatially periodic distortion of the extrudate [1].

In this paper, our focus is on mathematical modeling of the Lim-Schowalter experiment, in which the volumetric flow rate is prescribed by driving the fluid with a piston moving at fixed speed. Experimental data suggest that far away from the piston, the flow is essentially one-dimensional. Highly elastic and very viscous non-Newtonian materials are modeled mathematically as incompressible viscoelastic fluids with fading memory and, for the present, under isothermal conditions. The basis for our explanation of persistent oscillations in piston-driven flow is the differential Johnson-Segalman-Oldroyd constitutive law, which is consistent with experimental shear data; it exhibits a non-monotonic relation between steady shear stress and strain rate. This constitutive model has already been used successfully by us in Refs. [6, 7, 12] to provide an alternative explanation of "spurt" and related phenomena, originally observed in experiments by G. Vinogradov *et al.* [14] in pressure-gradient driven flows. (While the Johnson-Segalman-Oldroyd constitutive model performs well in reproducing particular shear flows of highly elastic and viscous non-Newtonian fluids, we neither claim nor expect that it will perform equally well in other flows (*e.g.*, extensional flow).)

Piston-driven flow at a fixed volumetric flow rate is modeled as an instantaneous, globally well-posed feedback-control problem in one space dimension, the control being the prescribed volumetric flow rate and the feedback being the pressure gradient. A deeper analytical understanding of the governing equations is possible because the system contains a small parameter, the ratio of Reynolds number to Deborah number. In the inertialess approximation, obtained by setting this parameter to zero, the equations governing the flow can be viewed as a continuous family of quadratic ordinary differential equations coupled by the non-local constraint that fixes the volumetric flow rate; all solutions of this system are bounded, even for large, discontinuous initial data. Numerical simulations demonstrate that beyond a critical flow rate, the time-asymptotic behavior is cyclic. Large shear strain rates are observed in a thin, but macroscopic, layer near the wall; such a layer also appears in pressure-driven spurt flow.

Further analysis of the inertialess system links the observed oscillations to the occurrence of a Hopf bifurcation that spawns a periodic orbit beyond a critical flow rate. To understand this behavior, we observe that the dynamic equations admit discontinuous steady states when the volumetric flow rate is fixed (corresponding to the wall stress between the steady local shear stress maximum and minimum), and we linearize the system around a discontinuous steady state with a single jump. Parametrizing such discontinuous steady states by wall stress and flow rate, there are regions in parameter space in which the eigenvalues of the discrete spectrum of the linearized operator change from having negative real parts to having positive real parts, and there is a separating curve along which the real part of the eigenvalue vanishes. This information is the basis for showing in Ref. [10] that a Hopf bifurcation to a periodic orbit occurs when the flow rate is raised beyond a critical value. However, proving that the resulting periodic orbit is stable (*i.e.*, that the bifurcation is supercritical) remains a challenging open problem.

The outline of this paper is as follows. In Sec. 2, we summarize the development the mathematical model for unsteady, isothermal, piston-driven planar channel flow of a highly elastic and viscous non-Newtonian fluid, based on balance laws and the Johnson-Segalman–Oldroyd differential constitutive law, with a single relaxation time, in three dimensions. The inertialess approximation for such a flow takes the form of a one-parameter quadratic system of first-order functional differential equations in which the fixed volumetric flow rate is imposed as the control and the pressure gradient is the feedback. In Sec. 3, a two-parameter family of discontinuous steady state flows is determined. The two parameters can be chosen to be the volumetric flow rate and the position of the stress discontinuity. In Sec. 4, we use a numerical algorithm for the inertialess system to simulate piston-driven flows. We consider the sudden start-up problem, in which the volumetric flow rate is instantaneously raised from zero and then held fixed. Upon varying the fixed volumetric flow rate, the numerical simulation of solutions exhibits transitions to and from a regime with persistent oscillations; this phenomenon compares favorably with the Lim-Schowalter observations. In Sec. 5, spectral properties of the inertialess system linearized about particular discontinuous steady states are determined. The infinite dimensional linearized operator is bounded, and we determine criteria for its essential and discrete spectra analytically. We use a numerical method to compute the discrete spectrum of the linearized operator from the analytical criterion. The discrete spectrum consists of a pair of simple, complex conjugate eigenvalues, and the sign of the real parts of these eigenvalues changes from negative to positive along a curve within the parameter plane for the discontinuous steady states. In Sec. 6, we relate the results on the discrete spectrum obtained in Sec. 5 to results of numerical simulations in Sec. 4, and we explain the experimentally observed persistent oscillations as a Hopf bifurcation occurring when the control parameter is increased beyond a critical value. Finally, in Sec. 7, we use the results of Secs. 5 and 6 to offer an alternative explanation of the mechanism underlying the Lim-Schowalter experiment.

2. Piston-Driven Shear Flow

To model the experiments of Lim and Schowalter, we consider the planar Poiseuille flow a channel of a highly elastic and very viscous incompressible non-Newtonian fluid under isothermal conditions, satisfying the Johnson-Segalman–Oldroyd constitutive relations for the evolution of the extra stress tensor. We summarize the derivation of the system of partial differential equations that governs such a one-dimensional shear flow, starting from balance laws and constitutive equations in three dimensions; for more details, see Refs. [6, 7].

The channel is aligned along the y-axis and extends between $x = -h/2$ and $x = h/2$. The flow is assumed to be symmetric about the centerline $x = 0$ of the channel, and we restrict attention to the interval $x \in [-h/2, 0]$. Since the fluid undergoes simple shearing, the velocity and stress variables are independent of y. In particular, the velocity field is $\mathbf{v} = (0, v(x,t))$, so that the flow is incompressible and the conservation of mass equation is automatically satisfied.

The total shear stress on the fluid is the sum of three contributions, an isotropic pressure, a Newtonian stress, and an extra stress $\boldsymbol{\pi}$. The conservation of momentum in the x-direction implies that the pressure takes the form $p = p_0(x,t) - f(t)\,y$, with f being the pressure gradient. We adopt the Johnson-Segalman–Oldroyd differential constitutive law with a single relaxation time [13, 2] to determine the extra stress. In shear flow, the extra stress is expressible in terms of two variables, the shear stress $\sigma(x,t) := \pi^{xy}$ and the principal normal stress difference $Z(x,t) := -\frac{1}{2}(1-a^2)(\pi^{yy} - \pi^{xx})$; here $a \in (-1,1)$ is the slip parameter of the fluid. The fluid variables v, σ, and Z are governed by the y-momentum equation and constitutive differential equations. We introduce two dimensionless parameters that characterize the flow: α, the ratio of Reynolds number to Deborah number; and ε, the ratio of Newtonian viscosity to shear viscosity. After suitable scaling, the governing equations become

$$\begin{aligned} \alpha v_t - \sigma_x &= \varepsilon v_{xx} + f\ , \\ \sigma_t - (Z+1)v_x &= -\sigma\ , \\ Z_t + \sigma v_x &= -Z\ . \end{aligned} \tag{JSO}$$

on the interval $[-1/2, 0]$. The boundary conditions are

$$v(-1/2,t) = 0 \quad \text{and} \quad v_x(0,t) = 0\ . \tag{BC}$$

At $t = 0$, we impose the initial conditions

$$v(x,0) = v_0(x)\ , \quad \sigma(x,0) = \sigma_0(x)\ , \quad \text{and} \quad Z(x,0) = Z_0(x)\ . \tag{IC}$$

For consistency of (IC) with (BC), we assume that $v_0(-1/2) = 0$ and $v_0'(0) = 0$; to maintain continuity at $x = 0$, we require that $\sigma_0(0) = 0$. This condition, together with the boundary condition $v_x(0,t) = 0$ and the second equation of system (JSO), implies that $\sigma(0,t) = 0$ for all t. Thus symmetry about the centerline in maintained.

In the case of highly elastic polymer systems, α is very small, 7 to 10 orders of magnitude smaller than ε, which itself is of the order of 10^{-3}. Throughout this paper, therefore, we make the inertialess approximation $\alpha = 0$. This approximation has been used successfully in Refs. [7, 12] to study pressure-driven flow in which the pressure gradient f is prescribed. If $\alpha = 0$, the momentum equation in system (JSO), combined with the boundary conditions (BC) and $\sigma(0,t) = 0$, implies that

$$v_x = \frac{T - \sigma}{\varepsilon}\ , \tag{2.1}$$

where

$$T(x,t) := -f(t)x \tag{2.2}$$

coincides with the total shear stress $\sigma + \varepsilon v_x$. Using Eq. (2.1), v_x can be eliminated from system (JSO).

In piston-driven flow, the volumetric flow rate Q is prescribed, Q being given by

$$Q(t) := 2\int_{-1/2}^{0} v(x,t)\,dx = -2\int_{-1/2}^{0} x v_x(x,t)\,dx\ ; \tag{2.3}$$

here an integration by parts was carried out in the second step. However, the pressure gradient f is not known; instead, f adjusts to maintain the desired flow rate Q. We can express f, or equivalently T, in terms of the given flow rate Q

as follows. Substituting for v_x in Eq. (2.3) using Eq. (2.1) leads to the feedback equation

$$T(x,t) = -12\varepsilon Q(t)x + 24x \int_{-1/2}^{0} y\sigma(y,t)\,dy\ , \tag{FB}$$

where we think of the prescribed Q as the control and f (or equivalently T) as the feedback. This equation is written more simply as $T = T_0 + P\sigma$, where $T_0(x,t) := -12\varepsilon Q(t)x$ and where the operator P, defined by

$$[P\sigma]\,(x,t) := 24x \int_{-1/2}^{0} y\sigma(y,t)\,dy \tag{2.4}$$

is a projection, in the sense that $P^2\sigma := P[P\sigma] = P\sigma$.

Thus the system governing inertialess piston-driven flow is the quadratic system of functional differential equations

$$\begin{aligned} \sigma_t &= (Z+1)\left(\tfrac{T-\sigma}{\varepsilon}\right) - \sigma\ , \\ Z_t &= -\sigma\left(\tfrac{T-\sigma}{\varepsilon}\right) - Z\ , \\ T &= T_0 + P\sigma\ , \end{aligned} \tag{QFDE}$$

together with prescribed the initial conditions (IC) for $\sigma(x,0), Z(x,0)$.

Throughout the rest of this paper, we will assume that the prescribed flow rate $Q(t) = \overline{Q}$ is a positive constant. It is shown in Ref. [9] that the initial value problem for $(QFDE)$ is globally well posed for all time and for initial data σ_0, Z_0 that are either smooth or rough (in the sense of being only bounded and measurable) and of arbitrary size; moreover, the stress components σ, Z remain bounded. (We remark that the global well-posedness of the full governing system for piston-driven flow with $\alpha > 0$ is also discussed in Ref. [9].) Beyond technical considerations that need not concern us here, the key to proving global existence and boundedness of solutions and understanding the dynamics is that solutions of system $(QFDE)$ satisfy the Lyapunov-type identity

$$\frac{d}{dt}\left\{\sigma^2 + (Z+1)^2\right\} = -2\left[\sigma^2 + (Z+\tfrac{1}{2})^2 - \tfrac{1}{4}\right]\ . \tag{L}$$

Observe that Eq. (L) is independent of α, ε, and T. It follows easily from Eq. (L) that $\sigma(\cdot,t)$ and $Z(\cdot,t)$ remain bounded for as long as the solution σ, Z of system $(QFDE)$ exists. This fact, together with standard results, implies that every locally defined solution can be continued forward (uniquely) for all time; moreover, the solution remains bounded, the bound depending only on the size of the initial data. However, although solutions to system $(QFDE)$ exist globally and remain bounded, it is difficult to determine their asymptotic behavior of solutions as $t \to \infty$. Indeed, the extensive literature on quadratic differential and functional differential equations does include any results on the global qualitative structure of solutions of $(QFDE)$. The rest of this paper is devoted to understanding some of the interesting dynamics of the quadratic system $(QFDE)$ by a combination of analytical and numerical methods, and this leads to an explanation of the experimental observations of Lim and Schowalter.

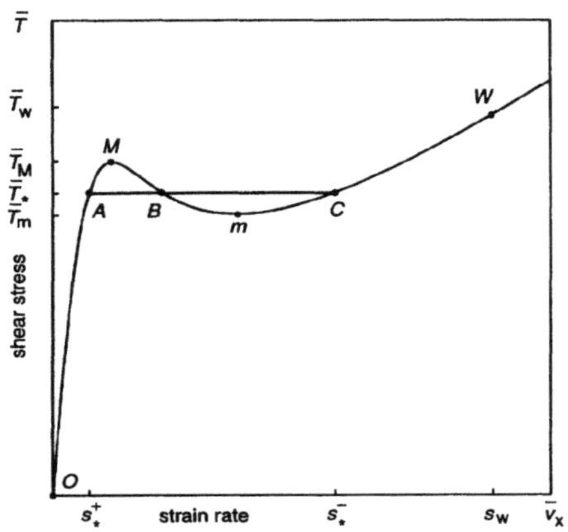

Fig. 1: The steady shear stress $\overline{T} = w(\overline{v}_x)$ plotted *vs.* shear strain rate $\overline{v}_x$.

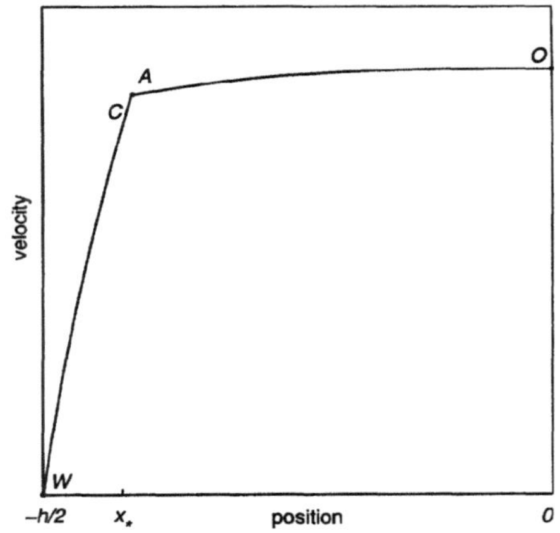

Fig. 2: The velocity profile in steady flow for the path $OACW$ shown in Fig. 1.

3. Steady Flows.

Our goal is to study the linearized stability and instability of steady state solutions of system $(QFDE)$. It will be clear that steady states of (JSO) and of $(QFDE)$ coincide. We shall indicate a steady state solution using an overbar. Taking $Q(t) := \overline{Q} > 0$ to be a constant and setting $\sigma_t = 0$ and $Z_t = 0$ in $(QFDE)$ shows that system $(QFDE)$ admits steady states $\overline{\sigma}, \overline{Z}$ satisfying the following requirements:

$$\overline{\sigma} = \frac{\overline{v}_x}{1+\overline{v}_x^2} \ , \qquad \overline{Z} = -\frac{\overline{v}_x^2}{1+\overline{v}_x^2} \ , \tag{3.1}$$

$$\overline{T} = \frac{\overline{v}_x}{1+\overline{v}_x^2} + \varepsilon \overline{v}_x \ , \qquad \overline{T} = \overline{T}_0 + P\overline{\sigma} \ , \tag{3.2}$$

where we define $\overline{T}_0(x) := -12\varepsilon\overline{Q}x$.

The steady strain rate $\overline{v}_x$ is determined by solving the steady stress/strain-rate relation

$$w(\overline{v}_x) = \overline{T} \ , \tag{3.3}$$

where the function w is defined by

$$w(s) := \frac{s}{1+s^2} + \varepsilon s \ . \tag{3.4}$$

When $\varepsilon < 1/8$, the function w is not monotone, as shown in Fig. 1. This leads to the occurrence of piecewise smooth steady state solutions with simple jump discontinuities in $\overline{v}_x$, implying a kink in the steady velocity $\overline{v}$; such a solution is indicated in Figs. 1 and 2. Formulae (3.1) imply that the steady stresses $\overline{\sigma}$ and $\overline{Z}$ are piecewise smooth functions having jump discontinuities whenever the strain rate $\overline{v}_x$ does.

While there are steady solutions with an arbitrary number of discontinuities, we focus on those with a single discontinuity. There is a two-parameter family of such solutions. It proves convenient to take as parameters the total stress $\overline{T}_* := \overline{T}(x_*)$ at the kink and the thickness $\delta := x_* + 1/2$ of the layer of high strain rate. The region of allowable values for these parameters is rectangular: $\overline{T}_*$ lies between $\overline{T}_m$ and $\overline{T}_M$, the stresses at the minimum and maximum of w; and δ lies in the interval $[0, 1/2)$.

For each choice of $(\overline{T}_*, \delta)$, the strain rate $\overline{v}_x$ is determined by solving

$$w(\overline{v}_x(x)) = \overline{T}_* \frac{x}{x_*} \tag{3.5}$$

for each $x \in [-1/2, 0]$; here $x_* := -1/2 + \delta$, and when there are multiple solutions of Eq. (3.5), $\overline{v}_x(x)$ is chosen on the high or low strain-rate branch of w according as $x < x_*$ or $x > x_*$. The function $\overline{v}_x$ determines $\overline{\sigma}$ and $\overline{Z}$ through Eqs. (3.1), and the flow-rate $\overline{Q}$ for the solution is calculated through Eq. (2.3).

A more convenient formula for $\overline{Q}$ can be obtained using a change of variables introduced by Yao [15]: in Eq. (2.3), change variables from x to $s := \overline{v}_x(x)$ using the relation $-\overline{f}x = w(s)$ to obtain

$$\overline{Q} = \frac{2}{\overline{f}^2} \int_{\mathcal{U}} s\, w(s) w'(s)\, ds \ , \tag{3.6}$$

where in the present application

$$\overline{f} = -\frac{\overline{T}_*}{x_*} = \frac{2\overline{T}_*}{1 - 2\delta} \ . \tag{3.7}$$

Here, the interval of integration is $\mathcal{U} := [0, s_*^+] \cup [s_*^-, s_w]$, where $s_*^\pm := \lim_{x \to x_*^\pm} \overline{v}_x(x)$ (the low and high strain-rate solutions of $w(s_*) = \overline{T}_*$) and $s_w := \overline{v}_x(-\frac{1}{2})$ (the high strain-rate solution of $w(s_w) = \overline{T}_w$), as in Fig. 1. The integral in Eq. (3.6) can be evaluated explicitly [15]:

$$\int_0^s t\, w(t) w'(t)\, dt = -\left(\tfrac{1}{4} - \varepsilon\right) \tan^{-1} s + \left(\tfrac{3}{4} - \varepsilon\right) \frac{s}{1+s^2} - \tfrac{1}{2} \frac{s}{(1+s^2)^2} + \tfrac{1}{3}\varepsilon^2 s^2 \ . \tag{3.8}$$

To make contact with experiment, however, it is preferable to use parameters different from $\overline{T}_*$ and δ, namely the total stress $\overline{T}_w := \overline{T}_*/(1 - 2\delta)$ at the channel wall and the flow rate $\overline{Q}$. (See Fig. 4a below.) The lines $\overline{T}_* = \overline{T}_m$, $\overline{T}_* = \overline{T}_M$, and $\delta = 0$, which bound the allowable region in the $(\overline{T}_*, \delta)$-plane, correspond respectively to the three curves labeled "bottom-jumping," "top-jumping," and "classical flow" in the $(\overline{T}_w, \overline{Q})$-plane.

4. Numerical Simulation of Piston-Driven Flows.

In order to motivate the analysis presented in subsequent sections, we first present results of numerically simulating the Lim-Schowalter experiment using system $(QFDE)$.

The numerical algorithm (which was derived in Ref. [9]) is as follows. We predict v_x at time level $n+1$ using the formula

$$(v_x)_{n+1} = \frac{T_n^k - \sigma_n^k}{\varepsilon} \tag{4.1}$$

for the kth element in the mesh; here k runs from 1 to the number of elements, N, and σ_n^k is the value of the shear stress in the kth element at time level n. For the given flow rate $\overline{Q}$, the stress T_n^k is computed by evaluating $P\sigma$ in system $(QFDE)$ using the midpoint rule:

$$T_n^k = \frac{(x_k + x_{k+1})}{2}\left(-12\varepsilon\overline{Q} + 24\sum_{i=1}^{N-1} \sigma_n^i\,(x_{i+1} - x_i)\,(x_i + x_{i+1})\,/2\right) . \tag{4.2}$$

We then difference and advance the stresses σ and Z, using the latest available value of $(v_x)_{n+1}$, by the following "corrector" scheme:

$$\begin{cases} \sigma_{n+1} = (1-\Delta t)\sigma_n + \Delta t(Z_n + 1)(v_x)_{n+1} \,, \\ Z_{n+1} = (1-\Delta t)Z_n - \Delta t\sigma_{n+1}(v_x)_{n+1} \,. \end{cases} \tag{4.3}$$

The basic simulation that we perform using the above algorithm idealizes the instantaneous start-up of a flow. In other words, the test apparatus is filled with fluid at rest, and then the flow rate is suddenly increased from zero to the prescribed value $\overline{Q}$. Therefore the initial conditions are $v_0(x) = \sigma_0(x) = Z_0(x) = 0$. To compare our results with the experimental results, we compute the pressure gradient f for each given flow rate $\overline{Q}$. In the simulations of $(QFDE)$, we observe four distinct flow regimes, corresponding to different values of the prescribed flow rate $\overline{Q}$: the classical, spurt I, oscillatory, and spurt II regimes. Typical plots of $f(t)$ *vs.* t in the four regimes are shown in Figs. 3a–d.

In the classical regime, continuous steady solutions are achieved. Steady solutions are also achieved in the spurt I and II regimes; they are, however, discontinuous and have the appearance of the spurt solutions achieved in pressure-gradient driven flows [6, 7]. As $\overline{Q}$ is increased and the transition to the oscillatory regime is approached, the time required for the flow to settle to a steady state increases, and the settling is accompanied by damped oscillations in $f(t)$[and in the velocity near the wall. Finally, at a critical value of $\overline{Q}$, the oscillations appear to become undamped and fail to settle down; this is the oscillatory regime. If $\overline{Q}$ is increased further, a second transition is observed, and steady, spurt-like solutions are again achieved; this is the spurt II regime. (For technical reasons, Fig. 3d terminates before the oscillations have died away to graphical accuracy; however, the simulation has been carried out for much longer time, and there is no doubt that the oscillations die out.)

We wish to stress that because we rely on numerical simulation, and because it is difficult to accurately distinguish slightly damped systems from undamped systems numerically, the existence of a true oscillatory regime for the full system $(QFDE)$ can only be conjectured at this point.

We studied transitions between these regimes in detail, as a function of the imposed volumetric flow rate $\overline{Q}$, for the polyisoprene sample PI-7 of Ref. [14] (*i.e.*, $\varepsilon = 0.001417$). As $\overline{Q}$ is increased from zero, the transition from the classical regime

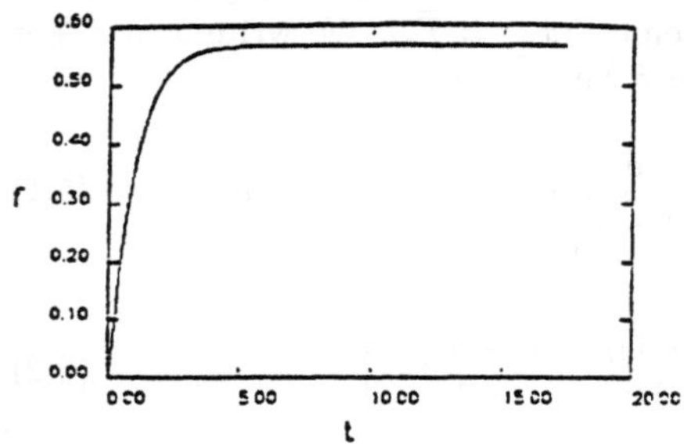

Fig. 3a: $f(t)$ *vs.* t in the classical regime of system ($QFDE$).

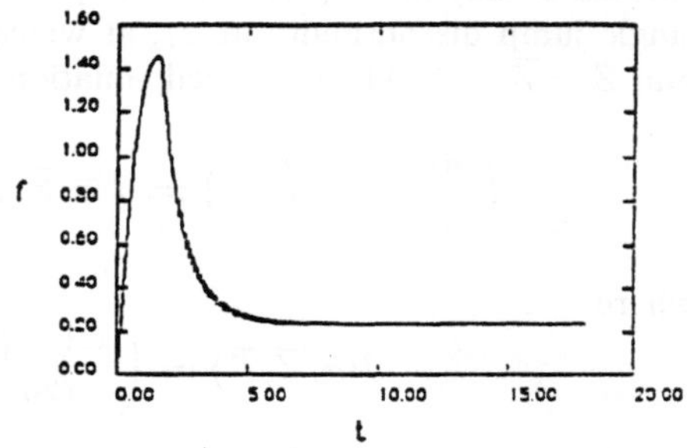

Fig. 3b: $f(t)$ *vs.* t in the spurt I regime of system ($QFDE$).

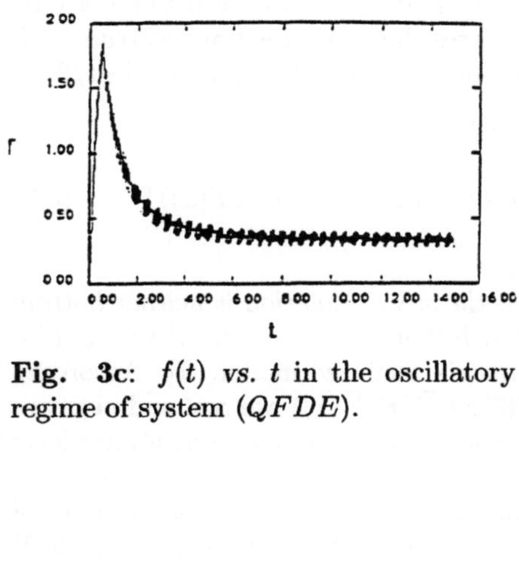

Fig. 3c: $f(t)$ *vs.* t in the oscillatory regime of system ($QFDE$).

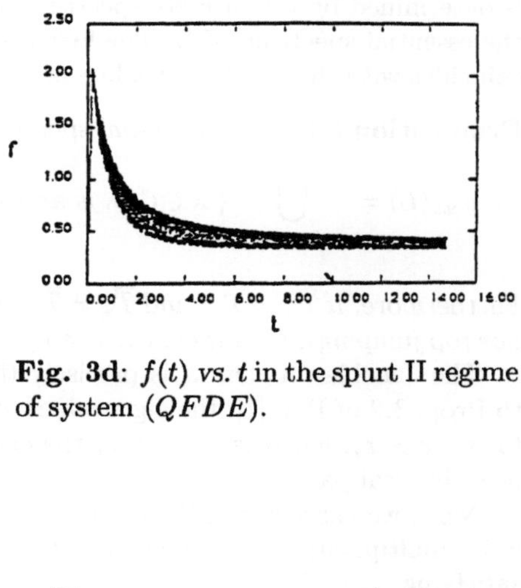

Fig. 3d: $f(t)$ *vs.* t in the spurt II regime of system ($QFDE$).

to the spurt I regime occurs at approximately $\overline{Q} = 0.1$. The transition between the steady spurt I regime and the oscillatory regime occurs at a critical $\overline{Q}$ of about 0.3, whereas the transition from the oscillatory regime back to the steady spurt II regime occurs at a critical $\overline{Q}$ of about 1.4. We emphasize that Lim [4] also observed four separate flow regimes in his experiments, the third of which is oscillatory and the fourth of which is a relatively steady regime at high shear rate (inferred from $\overline{Q}$). Thus our numerical simulation of ($QFDE$) conforms qualitatively with experimental results.

5. Linearized Stability Analysis.

We consider a solution of system $(QFDE)$ that is a small perturbation of a piecewise smooth steady state $(\overline{\sigma}, \overline{Z})$; for simplicity, we assume that this steady state has a single jump discontinuity at x_*, at which point $\overline{T}(x_*) = \overline{T}_*$. We write $\sigma = \overline{\sigma} + \hat{\sigma}$ and $Z = \overline{Z} + \hat{Z}$. The linearized equations for $\hat{\sigma}$ and $\hat{Z}$ are

$$\begin{pmatrix} \hat{\sigma}_t \\ \hat{Z}_t \end{pmatrix} = L \begin{pmatrix} \hat{\sigma} \\ \hat{Z} \end{pmatrix} := J(\overline{\sigma}, \overline{Z}, \overline{T}) \begin{pmatrix} \hat{\sigma} \\ \hat{Z} \end{pmatrix} + \begin{pmatrix} (\overline{Z}+1)/\varepsilon \\ -\overline{\sigma}/\varepsilon \end{pmatrix} P\hat{\sigma} , \tag{5.1}$$

where

$$J(\overline{\sigma}, \overline{Z}, \overline{T}) := \begin{pmatrix} -1 - (\overline{Z}+1)/\varepsilon & (\overline{T} - \overline{\sigma})/\varepsilon \\ (2\overline{\sigma} - \overline{T})/\varepsilon & -1 \end{pmatrix} . \tag{5.2}$$

The operator L is a bounded linear operator acting on the Hilbert space of pairs of functions $(\hat{\sigma}, \hat{Z}) \in X := PS_{x_*} \times PS_{x_*}$, where $PS_{x_*} := H^1(-1/2, x_*) \times H^1(x_*, 0)$ is the space of piecewise smooth (*i.e.*, H^1) functions on the interval $[-1/2, 0]$ with a single jump discontinuity at x_*. The linearized stability of the steady state is determined by the spectral properties of L which we now summarize; the results stated below are proved in Ref. [10].

Since the linear operator L is bounded on X, L is the generator of an analytic semigroup, and the time-asymptotic behavior of solutions of the linear system (5.1) is determined by the discrete spectrum of L. To find it, we also need to determine the essential spectrum of L. The first result states that the essential spectrum of L coincides with the set of eigenvalues of the Jacobian matrix J for every $x \in [-1/2, 0]$.

Proposition 5.1: *The essential spectrum of L is*

$$\sigma_{\rm ess}(L) = \bigcup_{x \in [-1/2, 0]} \{ \lambda \in C \mid \lambda \text{ is an eigenvalue of } J(\overline{\sigma}(x), \overline{Z}(x), \overline{T}(x)) \} . \tag{5.3}$$

Furthermore, if $\overline{T}_* \neq \overline{T}_m$ and $\overline{T}_* \neq \overline{T}_M$ (i.e., the steady solution is neither bottom nor top jumping), then there exists an $\eta > 0$ such that $\mathrm{Re}\,\lambda \leq -\eta$ for all $\lambda \in \sigma_{\rm ess}(L)$.

The Jacobian matrix J is precisely the matrix analyzed in Ref. [7]. According to Prop. 3.2 of Ref. [7], the eigenvalues of $J(\overline{\sigma}(x), \overline{Z}(x), \overline{T}(x))$ are real and negative for all $x > x_*$, whereas for $x < x_*$ the eigenvalues can be non-real, but always have negative real parts.

Next, we characterize the discrete spectrum of L, *i.e.*, its isolated eigenvalues of finite multiplicity. An element $\lambda \in \sigma_{\rm disc}(L)$ is such that there exists $(\tilde{\sigma}, \tilde{Z}) \neq (0,0)$ satisfying

$$\lambda \begin{pmatrix} \tilde{\sigma} \\ \tilde{Z} \end{pmatrix} = J(\overline{\sigma}, \overline{Z}, \overline{T}) \begin{pmatrix} \tilde{\sigma} \\ \tilde{Z} \end{pmatrix} + \begin{pmatrix} (\overline{Z}+1)/\varepsilon \\ -\overline{\sigma}/\varepsilon \end{pmatrix} P\tilde{\sigma} . \tag{5.4}$$

Row manipulations and use of the steady state relationships $\overline{\sigma} = (\overline{Z}+1)(\overline{T} - \overline{\sigma})/\varepsilon$ and $\overline{Z} = -\overline{\sigma}(\overline{T} - \overline{\sigma})/\varepsilon$ show that this condition is equivalent to

$$\begin{pmatrix} \lambda + 1 + (\overline{Z}+1)/\varepsilon & \overline{Z}/\overline{\sigma} \\ (\lambda+2)\overline{\sigma} & (\lambda+2)\overline{Z} + \lambda + 1 \end{pmatrix} \begin{pmatrix} \tilde{\sigma} \\ \tilde{Z} \end{pmatrix} = \begin{pmatrix} (\overline{Z}+1)/\varepsilon \\ 0 \end{pmatrix} P\tilde{\sigma} . \tag{5.5}$$

Since $\lambda \notin \sigma_{\text{ess}}(L)$, one can solve Eq. (5.5) for $(\tilde{\sigma}, \tilde{Z})$ in terms of $P\tilde{\sigma}$. The solution is

$$\begin{pmatrix} \tilde{\sigma} \\ \tilde{Z} \end{pmatrix} = \begin{pmatrix} \Sigma(\lambda, x) \\ \mathcal{Z}(\lambda, x) \end{pmatrix} P\tilde{\sigma} \,, \tag{5.6}$$

where

$$\Sigma(\lambda, x) := \frac{[(\lambda+2)\overline{Z} + \lambda + 1](\overline{Z}+1)/\varepsilon}{[(\lambda+2)\overline{Z} + \lambda + 1][\lambda + 1 + (\overline{Z}+1)/\varepsilon] - (\lambda+2)\overline{Z}} \,, \tag{5.7}$$

$$\mathcal{Z}(\lambda, x) := \frac{-(\lambda+2)\overline{\sigma}(\overline{Z}+1)/\varepsilon}{[(\lambda+2)\overline{Z} + \lambda + 1][\lambda + 1 + (\overline{Z}+1)/\varepsilon] - (\lambda+2)\overline{Z}} \,. \tag{5.8}$$

We emphasize that the denominator in these expressions vanishes for some $x \in [-\frac{1}{2}, 0]$ if and only if $\lambda \in \sigma_{\text{ess}}(L)$. The preceding calculations lead to the following characterization of the discrete spectrum:

Lemma 5.2: *A complex number λ belongs to $\sigma_{\text{disc}}(L)$ if and only if $\lambda \notin \sigma_{\text{ess}}(L)$ and*

$$24 \int_{-1/2}^{0} x^2 \Sigma(\lambda, x)\, dx = 1 \,. \tag{5.9}$$

Moreover, if $\lambda \in \sigma_{\text{disc}}(L)$, then λ has multiplicity 1.

The integral in the eigenvalue condition (5.9) can be evaluated by changing variables from x to $s := \overline{v}_x(x)$ using the relation $-\overline{f}x = w(s)$ (cf. Eq. (3.6)):

$$1 - 24 \int_{-1/2}^{0} x^2 \Sigma(\lambda, x)\, dx = \frac{24}{\overline{f}^3} \int_{\mathcal{U}} w(s)^2 \left[1 - \tilde{\Sigma}(\lambda, s)\right] w'(s)\, ds \,, \tag{5.10}$$

where the interval of integration is $\mathcal{U} := [0, s_*^+] \cup [s_*^-, s_w]$ and

$$\tilde{\Sigma}(\lambda, s) := \frac{\lambda + 1 - s^2}{(\lambda+1)\,[1 + \varepsilon(\lambda+1)] - [1 - \varepsilon - \varepsilon(\lambda+1)^2]\, s^2 + \varepsilon s^4} \,. \tag{5.11}$$

We define $\Phi(\lambda, \overline{T}_*, \delta)$ to be $\varepsilon\overline{f}^3/24$ times the right-hand side of Eq. (5.10), *i.e.*,

$$\Phi(\lambda, \overline{T}_*, \delta) := \int_{\mathcal{U}} \frac{s^2\,[(\lambda+1)^2 + s^2]\,\left[1 + \varepsilon + \varepsilon s^2\right]^2 \left[1 + \varepsilon - (1-2\varepsilon)s^2 + \varepsilon s^4\right]}{(1+s^2)^3 \left\{(\lambda+1)\,[1 + \varepsilon(\lambda+1)] - [1 - \varepsilon - \varepsilon(\lambda+1)^2]\, s^2 + \varepsilon s^4\right\}}\, ds \,. \tag{5.12}$$

The functions Φ depends on the parameters $\overline{T}_*$ and δ through s_*^-, s_*^+, and s_w, which determine $\mathcal{U}$. As an immediate consequence of Lemma 5.2, we have the following characterization of the discrete spectrum of L.

Proposition 5.3: *The discrete spectrum of L is*

$$\sigma_{\text{disc}}(L) = \left\{ \lambda \in \mathcal{C} \backslash \sigma_{\text{ess}}(L) \mid \Phi(\lambda, \overline{T}_*, \delta) = 0 \right\} \,. \tag{5.13}$$

While the integral in the definition of Φ can be evaluated explicitly, its exact form is not illuminating for computing the discrete spectrum analytically, and we determine it numerically as follows. A numerical scheme is used to solve the equation

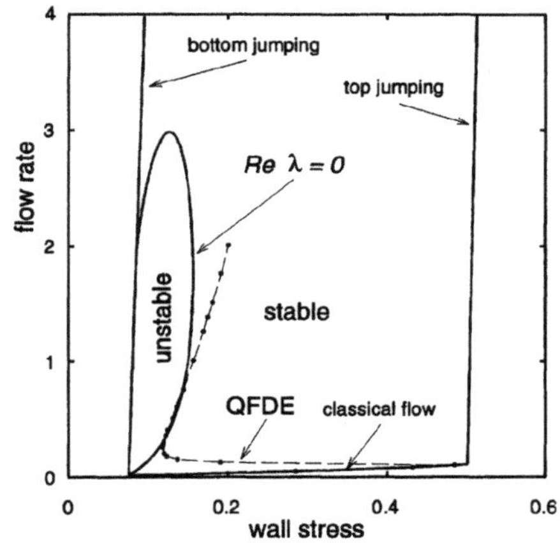

Fig. 4a: The regions of linearized stability and instability for steady state solutions with a single jump. The dots connected by dashed lines represent data from start-up simulations of system $(QFDE)$.

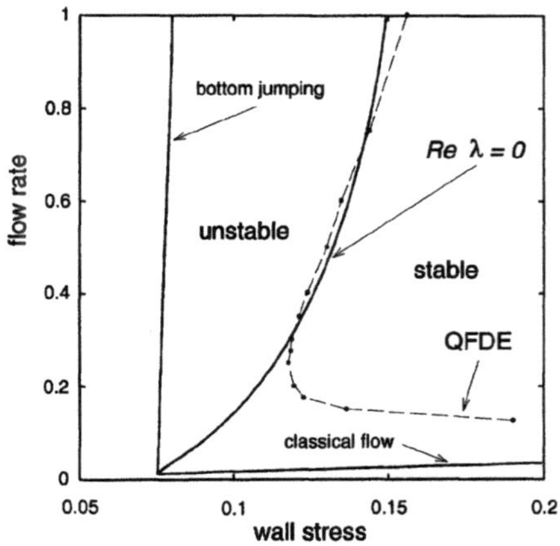

Fig. 4b: A closeup view of the portion of Fig. 6a where start-up simulations of system $(QFDE)$ leads to instability.

$\Phi(\lambda, \overline{T}_*, \delta) = 0$ for λ, given $\overline{T}_*$ and δ. This is a system of two real equations, $\operatorname{Re}\Phi(a+b\,i, \overline{T}_*, \delta) = 0$ and $\operatorname{Im}\Phi(a+b\,i, \overline{T}_*, \delta) = 0$, for the real and imaginary parts of $\lambda := a + bi$. We accounted for the possibility of having multiple roots by plotting the two level curves $\operatorname{Re}\Phi = 0$ and $\operatorname{Im}\Phi = 0$ in the (a, b)-plane using a contour plotter and looking for their intersections. For the range of parameters $\overline{T}_*$ and δ we explored, there is always exactly one intersection, and hence a single root.

We are interested in the possibility that this element λ of the discrete spectrum have $\operatorname{Re}\lambda > 0$, so that the corresponding steady state solution $(\overline{\sigma}, \overline{Z})$ is linearly unstable. Regarding λ now as a function of $\overline{T}_*$ and δ, instability would occur in a certain region of the $(\overline{T}_*, \delta)$-plane. The region in of instability is separated from the region of stability by the set of $(\overline{T}_*, \delta)$ such that $\operatorname{Re}\lambda = 0$. Thus we are led to find the curve of parameters $(\overline{T}_*, \delta)$ for which there exists a $b \in \mathbf{R}$ such that $\Phi(b\,i, \overline{T}_*, \delta) = 0$.

The results of solving this system of equations numerically is depicted in Fig. 4a. (The viscosity ratio is taken to be $\varepsilon = 0.001417$, which corresponds [3] to the polyisoprene sample PI-7 of Ref. [14].) Instead of plotting the $\operatorname{Re}\lambda = 0$ curve in the $(\overline{T}_*, \delta)$-plane, however, we have used as coordinates the wall stress $\overline{T}_w := \overline{T}_*/(1-2\delta)$ and the flow rate $\overline{Q}$, as given in Eq. (3.6).

6. Relation of Analysis to Simulation.

To compare the results from numerical simulation of $(QFDE)$ in Sec. 4 with the linearized stability analysis of Sec. 5, we observe that for each $\overline{Q} > 0$, the numerical solution of a start-up problem for $(QFDE)$ contains one parameter, δ, that approaches a constant value as time tends to infinity. Associated with this value of δ and of $\overline{Q}$, there is a unique steady state $(\overline{\sigma}(x), \overline{Z}(x))$ (with a single jump discontinuity) having layer thickness δ and flow rate $\overline{Q}$; this steady state might not be stable, in which case the solution does not approach this steady state, but rather seems to tend to a limit cycle (in the σ, Z phase plane) that in some sense "encircles" the unstable steady state. This steady state, be it linearly stable or unstable, can be computed by solving the equation

$$\overline{Q} = Q_{steady}(\overline{T}_*, \delta) \tag{6.1}$$

for $\overline{T}_*$, where $Q_{steady}(\overline{T}_*, \delta)$ is the function defined by the right side of Yao's formula for the flow rate, Eq. (3.6), in which the relevant parameters have been written in terms of $\overline{T}_*$ and δ, using relations of Sec. 3.

We have simulated a sequence of start-up problems, with increasing values of $\overline{Q}$, and computed $\overline{T}_*$ for each value of δ obtained. This gives a sequence of points in the $(\overline{T}_*, \delta)$-plane. The results for $\varepsilon = 0.001417$ are plotted as the discrete points in Figs. 4a and 4b; the equivalent coordinates $(\overline{T}_w, \overline{Q})$ have been used, as described at the end of Sec. 5. Figure 4b shows that the curve defined by this set of points crosses transversely into and out of the unstable region as $\overline{Q}$ increases. The two crossing points correspond to the transition points between the oscillatory regime and the spurt I and II regimes, respectively. Thus the numerical results suggest that the region of linearly unstable steady states is explored by the dynamical start-up experiments. We emphasize that although some steady state solutions of $(QFDE)$ are linearly unstable, all solutions of $(QFDE)$ are bounded for all time (see, Sec. 2).

A plausible explanation of the observed behavior, as well as of the correspondence between the solutions of $(QFDE)$ and the linearized stability analysis, is that for increasing values of the flow rate $\overline{Q}$, $(QFDE)$ is undergoing Hopf bifurcation upon crossing into and out of a regime of periodic orbits. We prove such a result in Ref. [10] by showing how the infinite-dimensional flow problem generated by system $(QFDE)$ on the Hilbert space X is reduced (using the center manifold theorem) to one for a two-dimensional vector field to which the classical Hopf bifurcation theorem can be applied (see [11], Theorem 1.13). However, confirming that these periodic orbits are stable (hence observable in a physical experiment), requires showing that the bifurcation to periodic orbits is supercritical, and this remains an open problem for the present.

It is interesting to note that the frequency of these periodic orbits seems to have physical significance: in Ref. [9], the frequency is observed to be proportional to $1/\varepsilon$; according to the dimensional analysis leading to system $(QFDE)$ and the assumptions made in fitting model parameters to material data [3], this observation translates into a prediction that the dimensional frequency of oscillation is independent of molecular weight in a sequence of experiments varying molecular weight at the same fixed $\overline{Q}$. This seems to be the case with Lim's experimental data [4] to a reasonable degree of accuracy.

7. Conclusion

We have presented a mathematical model aimed at explaining the experiments of Lim and Schowalter. Our model of piston-driven channel shear flow of a highly elastic and very viscous non-Newtonian fluid uses a constitutive model characterized by a non-monotonic relationship between steady shear stress and strain rate. We have described the reduction of the three-dimensional equations of motion and stress to approximating, one-dimensional systems that can be studied by a combination of numerical and analytic methods. The inertialess approximation results in a quadratic system of functional differential equations in which the prescribed volumetric flow rate is imposed by a feedback control. Numerical solution of this system exhibits four distinct flow regimes, as do the experiments of Lim and Schowalter. The third of these regimes exhibits persistent oscillations that compare favorably to the Lim-Schowalter observations.

If the governing system (JSO) and its inertial approximation ($QFDE$) model the observations of Lim and Schowalter correctly, then the details of the flow are different from what these experimentalists assumed. Rather than a stick-slip flow, the flow that we predict in the oscillatory regime has a thin "apparent slip" layer that exists for all time. The layer is unstable, in the sense that there are large and persistent time variations in the apparent slip velocity; these are associated with persistent oscillations in the pressure gradient that is controlled by fixing the flow rate $\overline{Q}$. For a certain range of $\overline{Q}$, we have provided convincing evidence that these oscillations are a consequence of a Hopf bifurcation to periodic orbits by a combination of analytical and computational methods. Moreover, the frequency of these oscillations agrees with observations of Lim. Our earlier analysis of the "spurt phenomenon" in Refs. [7, 12] explains why the control is the source of the instability: the flow would become steady if the control were removed by prescribing the pressure gradient (although, of course, the desired flow rate would then almost surely not be achieved).

References

1. M. Denn, "Issues in viscoelastic fluid dynamics," *Ann. Rev. Fluid Mech.* **22** (1989), 13–34.
2. M. Johnson and D. Segalman, "A Model for Viscoelastic Fluid Behavior which Allows Non-Affine Deformation," *J. Non-Newtonian Fluid Mech.* **2** (1977), 255–270.
3. R. W. Kolkka, D. S. Malkus, M. G. Hansen, G. R. Ierley, and R. A. Worthing, "Spurt Phenomena of the Johnson-Segalman Fluid and Related Models," *J. Non-Newtonian Fluid Mech.* **29** (1989), 303–325.
4. F. J. Lim, "Wall slip in narrow molecular weight distribution polybutadienes," Ph.D. Thesis, Princeton Univ., 1988.
5. F. J. Lim and W. R. Schowalter, "Wall Slip of Narrow Molecular Weight Distribution Polybutadienes," *J. Rheology* **33** (1989), 1359.
6. D. S. Malkus, J. A. Nohel, and B. J. Plohr, "Dynamics of Shear Flow of a Non-Newtonian Fluid," *J. Comp. Phys.* **87** (1990), 464–487.
7. D. S. Malkus, J. A. Nohel, and B. J. Plohr, "Analysis of New Phenomena in Shear Flow of Non-Newtonian Fluids," *SIAM J. Appl. Math.* **51** (1991), 899–929.
8. D. S. Malkus, Y.-C. Tsai, and R. W. Kolkka, "New transient algorithms for non-Newtonian flows," *Finite Elements in Fluids* **8** (1992), 401–424.
9. D. S. Malkus, J. A. Nohel and B. J. Plohr, "Approximation of piston-driven flows of a non-Newtonian fluid," in *Control Theory, Dynamical Systems, and Geometry of Dynamics*, (1993),

Elworthy, Everitt, and Lee, Eds., Marcel Dekker, 173-192.

10. D. S. Malkus, J. A. Nohel and B. J. Plohr, "Piston-driven shear flow of a non-Newtonian fluid," in preparation.
11. J. E. Marsden and M. McCracken, *The Hopf Bifurcation and Its Applications*, Applied Mathematical Sciences 19, Springer-Verlag, New York, 1976.
12. J. Nohel and R. Pego, "Nonlinear Stability and Asymptotic Behavior of Shearing Motions of a Non-Newtonian Fluid," *SIAM J. Math. Anal.* **24** (1993), 911–942.
13. J. Oldroyd, "Non-Newtonian Effects in Steady Motion of Some Idealized Elastico-Viscous Liquids," *Proc. Royal Soc. London, Series A* **245** (1958), 278–297.
14. G. V. Vinogradov, A. Ya. Malkin, Yu. G. Yanovskii, E. K. Borisenkova, B. V. Yarlykov, and G. V. Berezhnaya, "Viscoelastic Properties and Flow of Narrow Distribution Polybutadienes and Polyisoprenes," *J. Polymer Sci., Part A-2* **10** (1972), 1061–1084.
15. M. Yao, "A Numerical and Analytical Study of Normal Stresses and Pressure Differences in non-Newtonian Creeping Flows," Ph.D. Thesis, Univ. of Wisconsin–Madison, 1989.

Department of Engineering Mechanics and the Center for the Mathematical Sciences, University of Wisconsin, Madison, WI 53706

E-mail: malkus@cms.wisc.edu

Forschungsinstitut für Mathematik, ETH-Zentrum, CH 8092 Zürich, Switzerland, and Department of Mathematics and the Center for the Mathematical Sciences, University of Wisconsin, Madison, WI 53706

E-mail: jan@math.ethz.ch

Departments of Mathematics and of Applied Mathematics and Statistics, State University of New York at Stony Brook, Stony Brook, NY 11794-3651

E-mail: plohr@ams.sunysb.edu

SOME EXISTENCE AND STABILITY RESULTS FOR NON-ISOTHERMAL FLOWS OF VISCOELASTIC FLUIDS

V. BAZIN
Office National d'Etudes et de Recherches Aéronautiques
Le Fort de Palaiseau 91120 Palaiseau - FRANCE

C. GUILLOPÉ*
Mathématiques, UFR de Sciences et Technologie
Université Paris XII - Val de Marne, 61 avenue du Général de Gaulle
94010 Créteil Cedex - FRANCE

and

J.-C. SAUT
Laboratoire d'Analyse Numérique
Université Paris-Sud and CNRS, Bâtiment 425, 91405 Orsay Cedex - FRANCE

Abstract. In this talk we review some mathematical results obtained for non-isothermal flows of viscoelastic fluids having a differential constitutive equation. The effects of temperature changes in the parameters characterizing the fluid could be important in practical applications. We consider here the case of the Bénard problem where a viscoelastic fluid is placed in a container heated from below as well as the effect of viscous heating on the flow.

Key words: Bénard problem, Viscous heating, Differential constitutive equations

1. Introduction

The purpose of this study is to give, through mathematical analysis, some insight into the problem of instabilities observed in experiments with polymers. In non-isothermal flows of viscoelastic fluids two kinds of instabilities can be seen. The first one, the *hydrodynamic instability*, is well known and well understood for Newtonian fluids. Experiments such as Bénard convection in a container filled with a Newtonian liquid and heated from below are explained by the mathematical theory of bifurcation. A second type of instability is the *purely elastic instability* occuring at zero Reynolds number. It has been observed numerically in Couette-Taylor flows for example. (See Larson *et al.*, 1990.)

The viscoelastic fluids that are considered here are incompressible and satisfy a differential constitutive law. The stress tensor $\boldsymbol{\sigma}$ is given by

$$\boldsymbol{\sigma} = -pI + 2\eta_s \mathbf{D} + \boldsymbol{\tau},$$

where p denotes the hydrostatic pressure, $\mathbf{D} = \mathbf{D}[\mathbf{u}] = \frac{1}{2}(\nabla \mathbf{u} + \nabla \mathbf{u}^T)$ the rate of deformation tensor, and η_s the Newtonian viscosity. The extra-stress tensor $\boldsymbol{\tau}$ will

* And Laboratoire d'Analyse Numérique, CNRS and Université Paris-Sud.

J.F. Dijksman and G.D.C. Kuiken (eds.),
IUTAM Symposium on Numerical Simulation of Non-isothermal Flow of Viscoelastic Liquids, 73-84.

be given by a differential equation, which in its simplest form reads as follows

$$\boldsymbol{\tau} + \lambda \frac{\mathcal{D}_a \boldsymbol{\tau}}{\mathcal{D}t} = 2\eta_p \mathbf{D}.$$

The positive parameter η_p denotes the polymeric viscosity, and $\frac{\mathcal{D}_a}{\mathcal{D}t}$ the frame indifferent time derivative of a tensor:

$$\frac{\mathcal{D}_a \boldsymbol{\tau}}{\mathcal{D}t} = \frac{d\boldsymbol{\tau}}{dt} + \boldsymbol{\tau}\mathbf{W} - \mathbf{W}\boldsymbol{\tau} - a(\mathbf{D}\boldsymbol{\tau} + \boldsymbol{\tau}\mathbf{D}),$$

where $\frac{d}{dt} = \frac{\partial}{\partial t} + \mathbf{u} \cdot \nabla$ is the convected derivative, $\mathbf{W} = \frac{1}{2}(\nabla \mathbf{u} - \nabla \mathbf{u}^T)$ is the vorticity tensor, and a a real parameter, $-1. \leq a \leq 1$.

The remaining of this paper is organized as follows. Section 2 and 3 are devoted to the study of the Bénard problem. Section 2 deals with unsteady solutions: we show in particular that the pure conductive state is asymptotically Liapunov stable. In Section 3 we show the existence of slightly viscoelastic non-trivial states and study their stability. Finally Section 4 is devoted to the study of the effects on the flow of viscous heating, in particular in the case of one-dimensional flows.

Notation. Let Ω be a bounded domain of $\mathbf{R}^N$, $N = 1, 2,$ or 3. We introduce the following function spaces. $L^2(\Omega)$ is the space of square integrable functions on Ω, and $\mathbf{L}^2(\Omega)$ the space of vector valued functions with components in $L^2(\Omega)$. The norm in both these Hilbert spaces is denoted by $|\cdot|$. $H^m(\Omega)$, $m \in \mathbf{N}$, is the Sobolev space of order m, space of functions which have distributional derivatives of order up to m in $L^2(\Omega)$. $\mathbf{H}^m(\Omega)$ is the corresponding vector valued space, $\mathbf{H}^m_{\text{sym}}(\Omega)$ the corresponding space of tensor valued functions which are symmetric. We also introduce the space of functions which satisfy a periodicity condition on part of the boundary of Ω: these are denoted by $L^2_{\#}$, and $H^m_{\#}$.

We introduce the spaces of solenoidal vector functions: H is the subspace of $\mathbf{L}^2(\Omega)$ of such vector functions (with possibly periodic normal traces on part of the boundary), and V the subspace of $\mathbf{H}^1(\Omega)$ with zero boundary values and/or boundary conditions on part of the boundary. P denotes the orthogonal projection of $\mathbf{L}^2(\Omega)$ onto H, and A the Stokes operator: this is an isomorphism of V onto its dual space, as well as of $D(A) = \mathbf{H}^2(\Omega) \cap V$ onto H.

Finally we introduce the spaces of functions defined from a time interval $(0, T)$ into a functional space X: $L^p(0, T; X)$, for $p > 0$, possibly $p = \infty$, is the space of functions for which $\|u(t, \cdot)\|_X$ is in $L^p(0, T)$; $\mathcal{C}([0, T]; X)$ is the space of functions for which $t \longrightarrow u(t)$ is continuous from $[0, T]$ with values in X, and $\mathcal{C}_b([0, T]; X)$ the subspace of continuous functions which are bounded on $[0, T]$.

2. Convection of a Viscoelastic Flow Heated from Below Part I

We consider here a viscoelastic fluid between two parallel plates maintained at constant temperatures: the bottom plate $\{z = 0\}$ is at temperature $T_0 + \delta\mathrm{T} > T_0$, while the top one $\{z = d\}$ is at temperature T_0. The density is given by the Boussinesq approximation as a function of temperature:

$$\rho = \rho_0 \big(1 - \beta_v (T - T_0)\big),$$

where ρ_0 is the density at temperature T_0, and $\beta_v > 0$ the coefficient of volumetric thermal expansion.

We introduce the following non-dimensional numbers: the Prandtl number, $\mathrm{Pr} = \frac{\eta}{\rho_0 k}$, which measures the viscous effects compared to the conductivity effects; the Rayleigh number, $\mathrm{Ra} = \frac{\beta_v g d^3 \rho_0 \delta \mathrm{T}}{\eta k}$, which measures the effects due the difference of temperature $\delta\mathrm{T}$ of the plates compared to the viscous effects; the Deborah number, $\mathrm{De} = \frac{k\lambda_1}{d^2}$, which measures the elasticity of the liquid; and the retardation parameter, $\omega = \frac{\eta_p}{\eta}$, $0 < \omega < 1$, which is the ratio of the polymeric viscosity to the total viscosity $\eta = \eta_s + \eta_p$. Above k denotes the constant thermal conductivity, and g the magnitude of gravity.

We consider perturbations of the pure conduction state:

$$\overline{\mathbf{u}} = \mathbf{0}, \qquad \overline{\boldsymbol{\tau}} = \mathbf{0}, \qquad \overline{T} = T_0 + \frac{\delta \mathrm{T}}{d}\, z,$$

with the corresponding density $\overline{\rho}$, linear in $\overline{T}$, and the corresponding pressure $\overline{p}$, quadratic in $\overline{T}$. The non-dimensional equations for the perturbation $(\mathbf{u}, p, \boldsymbol{\tau}, \theta)$ are satisfied in the domain $\Omega = [0, L_1) \times [0, L_2) \times (0,1)$:

$$(\mathrm{Pr})^{-1} \frac{d\mathbf{u}}{dt} + \nabla p = (1-\omega)\Delta \mathbf{u} + \mathbf{div}\, \boldsymbol{\tau} + \mathrm{Ra}\, \theta \mathbf{k}, \tag{1}$$

$$\operatorname{div} \mathbf{u} = 0, \tag{2}$$

$$\mathrm{De} \frac{\mathcal{D}_a \boldsymbol{\tau}}{\mathcal{D}t} + \boldsymbol{\tau} = 2\omega \mathbf{D}, \tag{3}$$

$$\frac{d\theta}{dt} - w = \Delta \theta, \tag{4}$$

together with boundary conditions and initial conditions for $\mathbf{u}$, $\boldsymbol{\tau}$, and θ. Homogeneous boundary conditions are assumed for $\mathbf{u}$ and θ on the plates

$$\mathbf{u} = \mathbf{0}, \quad \theta = 0 \quad \text{on} \quad \{z = 0\} \quad \text{and} \quad \{z = 1\}. \tag{5}$$

Moreover all the functions and their derivatives up to an order depending on the regularity satisfy periodicity conditions on $\{x = 0\}$ and $\{x = L_1\}$ in one hand, and on $\{y = 0\}$ and $\{y = L_2\}$ in the other hand. The initial conditions are given on Ω,

$$\mathbf{u}(\cdot, 0) = \mathbf{u}_0, \quad \boldsymbol{\tau}(\cdot, 0) = \boldsymbol{\tau}_0, \quad \theta(\cdot, 0) = \theta_0. \tag{6}$$

A point of Ω is denoted by $\mathbf{x} = (x, y, z)$. In equation (1) w denotes the z-component of velocity, and in equation (4) $\mathbf{k}$ is the unit vector in the z-direction.

Using a fixed point theorem applied to a linearized problem of Problem (1)–(6), we first show the local existence of transient solutions. (See Bazin, 1990.)

Theorem 2.1 (Local existence of regular solutions)

Let $\mathbf{u}_0 \in D(A)$, $\boldsymbol{\tau}_0 \in \mathbf{H}^2_{\mathrm{sym}}(\Omega)$, and $\theta_0 \in H^1_{\#}(\Omega)$. Then there exists a time T^* and a solution $(\mathbf{u}, p, \boldsymbol{\tau}, \theta)$ to Probem (1)–(6) such that

$$\mathbf{u} \in L^2(O, T^*; \mathbf{H}^3(\Omega)) \cap \mathcal{C}([O, T^*]; D(A)), \quad \mathbf{u}' \in L^2(O, T^*; V) \cap \mathcal{C}([O, T^*]; H),$$

$$p \in L^2(O, T^*; H^2(\Omega)),$$

$$\boldsymbol{\tau} \in \mathcal{C}([O, T^*]; \mathbf{H}^2_{\text{sym}}(\Omega)), \quad \boldsymbol{\tau}' \in \mathcal{C}([O, T^*]; \mathbf{H}^1_{\text{sym}}(\Omega)),$$

$$\theta \in \mathcal{C}([O, T^*]; H^1_{\#}(\Omega)), \quad \theta' \in \mathcal{C}([O, T^*]; L^2_{\#}(\Omega)).$$

This solution is unique, the pressure being defined up to an additive constant.

Using some energy type estimates for $\mathbf{u}$, $\boldsymbol{\tau}$, and θ, we show that the solution exists for all time if the data are small enough, and is bounded in time.

Theorem 2.2 (Global existence of regular solutions for small initial data) *There exists a parameter ω_0, $0 < \omega_0 < 1$, and a Rayleigh number $\mathrm{Ra}_0 > 0$ such that, if $\omega \in (0, \omega_0]$ and $\mathrm{Ra} \in (0, \mathrm{Ra}_0]$, then for regular initial data $\mathbf{u}_0, \boldsymbol{\tau}_0, \theta_0$ small enough, the solution obtained in Theorem 2.1 is defined for all times. Precisely,*

$$\mathbf{u} \in \mathcal{C}_b(\mathbf{R}_+; D(A)) \cap L^2_{\text{loc}}(\mathbf{R}_+; \mathbf{H}^3(\Omega)), \quad \mathbf{u}' \in \mathcal{C}_b(\mathbf{R}_+; H) \cap L^2_{\text{loc}}(\mathbf{R}_+; V),$$

$$\boldsymbol{\tau} \in \mathcal{C}_b(\mathbf{R}_+; \mathbf{H}^2_{\text{sym}}(\Omega)), \quad \boldsymbol{\tau}' \in \mathcal{C}_b(\mathbf{R}_+; \mathbf{L}^2_{\text{sym}}(\Omega)),$$

$$\theta \in \mathcal{C}_b(\mathbf{R}_+; H^1_{\#}(\Omega)), \quad \theta' \in \mathcal{C}_b(\mathbf{R}_+; L^2_{\#}(\Omega)).$$

We also deduce immediately that the pure conductive steady solution $(\overline{\mathbf{u}}, \overline{p}, \overline{\boldsymbol{\tau}}, \overline{T})$ is asymptotically stable.

Theorem 2.3 (Stability of the pure conductive state)
Under the hypotheses of Theorem 2.2, the steady solution $(\mathbf{u}, \boldsymbol{\tau}, \theta) = (\mathbf{0}, \mathbf{0}, 0)$ to Problem (1)–(6) is asymptotically Liapunov stable in $\mathbf{H}^1(\Omega) \times \mathbf{H}^1_{\text{sym}}(\Omega) \times H^1_{\#}(\Omega)$.

3. Convection of a Viscoelastic Flow Heated from Below Part II

In this section we suppose that ω is small and we look for steady solutions to Problem (1)–(5) which are close to a Newtonian flow. We are going to show that for We and ω small enough, the Bénard problem admits a nontrivial solution which is linearly stable.

As in (Kirchgässner, 1975) we introduce the non-dimensional Grashof number, $\mathrm{Gr} = \frac{\mathrm{Ra}}{\mathrm{Pr}}$, and the bifurcation parameter $\mu = \sqrt{\mathrm{Gr}}$.

We also rescale the functions $\mathbf{u}$, $\boldsymbol{\tau}$, and p, by setting

$$\mathbf{u} = \mu \mathbf{v}, \qquad \boldsymbol{\tau} = \mu \boldsymbol{\sigma}, \qquad p = \mu q,$$

and we obtain the following boundary value problem deduced from the steady version of (1)–(5):

$$\mu(\mathbf{v} \cdot \nabla)\mathbf{v} - (1 - \omega)\Delta \mathbf{v} - \mu \theta \mathbf{k} + \nabla q = \mathbf{div}\, \boldsymbol{\sigma}, \tag{7}$$

$$\operatorname{div} \mathbf{v} = 0, \tag{8}$$

$$\mu \mathrm{De}\Big((\mathbf{v} \cdot \nabla)\boldsymbol{\sigma} + \boldsymbol{\beta}(\nabla \mathbf{v}, \boldsymbol{\sigma})\Big) + \boldsymbol{\sigma} = 2\omega \mathbf{D}, \tag{9}$$

$$\mu(\mathbf{v}\cdot\nabla)\theta - \mathrm{Pr}^{-1}\Delta\theta - \mu w = 0, \tag{10}$$

together with boundary conditions

$$\mathbf{v} = \mathbf{0}, \qquad \theta = 0 \qquad \text{on} \quad \{z=0\} \quad \text{.and} \quad \{z=1\}, \tag{11}$$

and the usual periodic boundary conditions on the sides $\{x = 0\}$ and $\{x = L_1\}$ on one hand, and $\{y = 0\}$ and $\{y = L_2\}$ on the other hand. Here $\boldsymbol{\beta}$ denotes a tensor valued function, at least quadratic in $\nabla\mathbf{u}$ and $\boldsymbol{\sigma}$, which contains for example the extra terms in the derivative $\frac{\mathcal{D}_a}{\mathcal{D}t}$. Here w is the z-component of $\mathbf{v}$.

The Newtonian case which corresponds to equations (7), (8), (10), and (11), where $\boldsymbol{\sigma} = \mathbf{0}$, has been studied in particular by Joseph, Kirchgässner, Iooss, and Iudovich. The results are the following.

Theorem 3.1 (Existence of non-trivial steady Newtonian states)
(i) (Joseph, 1976) *The conductive (trivial) solution* $(\mathbf{v},\theta) = (\mathbf{0},0)$ *of Problem (7), (8), (10), and (11) is unique for* μ *in the interval* $[0,\mu_1)$, *where* μ_1 *is the smallest positive characteristic value of the linearized operator at* $(\mathbf{0},0)$. *Moreover, if* μ_1 *is simple, the conductive solution is unique even for* $\mu = \mu_1$.

(ii) (Iudovich, 1966 and 1967) *Suppose that* μ_1 *is simple. Then there exists a unique nontrivial branch of solutions* $\big(\mathbf{v}(\delta),\theta(\delta)\big)$ *emanating from* $(\mathbf{v},\theta,\mu) = (\mathbf{0},0,\mu_1)$ *where* $\delta = \sqrt{\mu-\mu_1}$. *Moreover,*

$$\lim_{\delta\to 0}\frac{\left|\big(\mathbf{v}(\delta),\theta(\delta)\big)\right|}{\delta} = 1.$$

(iii) (Iudovich, 1966 and 1967) μ_1 *is simple for almost all* $\gamma^2 = L_1^2 + L_2^2$. *(More precisely, the multiplicity of* μ_1 *is one except for* L_1 *and* L_2 *in a countable set.)*

(iv) (Iooss, 1971, Iudovich, 1966 and 1967, and Kirchgässner, 1975) *Suppose that* μ_1 *is simple. Then the branching solution* $\big(\mathbf{v}(\delta),\theta(\delta)\big)$ *is asymptotically (Liapunov) stable in the space* $\mathbf{H}^2$ *(hence pointwise), for* μ $(> \mu_1)$ *close to* μ_1. *The basic solution is stable for* $0 \le \mu \le \mu_1$ *and unstable for* $\mu > \mu_1$.

For $\mu > \mu_1$ fixed, let $(\widetilde{\mathbf{v}}^{\mu},\widetilde{\theta}^{\mu},\widetilde{q}^{\mu})$ be a nontrivial solution of the (classical) Bénard problem,

$$\mu(\widetilde{\mathbf{v}}\cdot\nabla)\widetilde{\mathbf{v}} - \Delta\widetilde{\mathbf{v}} - \mu\widetilde{\theta}\mathbf{k} + \nabla\widetilde{q} = \mathbf{0}, \tag{12}$$

$$\operatorname{div}\widetilde{\mathbf{v}} = 0, \tag{13}$$

$$\mu(\widetilde{\mathbf{v}}\cdot\nabla)\widetilde{\theta} - \mathrm{Pr}^{-1}\Delta\widetilde{\theta} - \mu\widetilde{w} = 0, \tag{14}$$

together with boundary conditions (11) and with the periodic boundary conditions. In equation (14), $\widetilde{w}$ denotes the z–component of $\widetilde{\mathbf{v}}$.

We define the linear operator $L_\mu[\widetilde{\mathbf{v}}^{\mu},\widetilde{\theta}^{\mu}] : D(A)\times(H^2\cap H^1_{\#}) \longrightarrow H\times L^2$, by

$$L_\mu\Big[\widetilde{\mathbf{v}}^{\mu},\widetilde{\theta}^{\mu}\Big](\mathbf{v},\theta) = \Big(\mu P\Big\{(\widetilde{\mathbf{v}}^{\mu}\cdot\nabla)\mathbf{v} + (\mathbf{v}\cdot\nabla)\widetilde{\mathbf{v}}^{\mu}\Big\} + A\mathbf{v},$$
$$\mu\Big\{(\widetilde{\mathbf{v}}^{\mu}\cdot\nabla)\theta + (\mathbf{v}\cdot\nabla)\widetilde{\theta}^{\mu}\Big\} - \mathrm{Pr}^{-1}\Delta\theta - \mu w\Big). \tag{15}$$

Using an iterative scheme built from the operator $L_\mu[\widetilde{\mathbf{v}}^\mu, \widetilde{\theta}^\mu]$ in the same way as in (Renardy, 1987) we first show the existence of steady states. (See also Guillopé *et al.*, 1991.)

Theorem 3.2 (Existence of a non-trivial viscoelastic steady state)
Assume that $L_\mu\left[\widetilde{\mathbf{v}}^\mu, \widetilde{\theta}^\mu\right]$ *is an isomorphism and that*

$$C_\beta \mathrm{We}\, \|\widetilde{\mathbf{v}}^\mu\|_{\mathbf{H}^3} < 1, \tag{16}$$

where C_β *is a constant depending on the tensor valued function* $\boldsymbol{\beta}$. *Then there exists* ω_0, $0 < \omega_0 < 1$, *such that, for any* ω, $0 < \omega \leq \omega_0$, *the system (7)–(11) admits a unique solution* $(\mathbf{v}^\mu_\omega, \theta^\mu_\omega, \boldsymbol{\sigma}^\mu_\omega)$ *in* $\mathbf{H}^3 \cap D(A) \times H^3 \times \mathbf{H}^2_{\mathrm{sym}}$ *which tends to* $(\widetilde{\mathbf{v}}^\mu, \widetilde{\theta}^\mu, \mathbf{0})$ *as* $\omega \to 0$.

Remark 3.1 The operator $L_\mu\left[\widetilde{\mathbf{v}}^\mu, \widetilde{\theta}^\mu\right]$ is an isomorphism in particular if $\left(\widetilde{\mathbf{v}}^\mu, \widetilde{\theta}^\mu, \mu\right)$ is a solution given by Theorem 3.1, which is on the branch emanating from $(\mathbf{v}, \theta, \mu) = (\mathbf{0}, 0, \mu_1)$ for μ close to μ_1, μ_1 being simple.

Concerning the stability of the solution $(\mathbf{v}^\mu_\omega, \theta^\mu_\omega, \boldsymbol{\sigma}^\mu_\omega)$, we have the following result.

Theorem 3.3 (Linear stability of non-trivial viscoelastic steady states)
We assume that μ_1 *is simple and that* $\mu > \mu_1$ *is close enough to* μ_1. *Suppose that the corresponding Newtonian solution* $(\widetilde{\mathbf{v}}^\mu, \widetilde{\theta}^\mu)$ *satisfies condition (16). Then the viscoelastic solution* $(\mathbf{v}^\mu_\omega, \theta^\mu_\omega, \boldsymbol{\sigma}^\mu_\omega)$ *constructed in Theorem 3.2 is linearly asymptotically stable.*

For the proof of this result, we note that by Theorem 3.1 (iv), $(\widetilde{\mathbf{v}}^\mu, \widetilde{\theta}^\mu)$ is asymptotically stable. This, together with an easy extension of Prodi's result (1962), implies that the spectrum of $L_\mu[\widetilde{\mathbf{v}}^\mu, \widetilde{\theta}^\mu]$ can be localized in a parabola like set. Moreover, we can also localize the spectrum of the linearized operator around $(\mathbf{v}^\mu_\omega, \theta^\mu_\omega, \boldsymbol{\sigma}^\mu_\omega)$ and prove an estimate of the resolvent. The linear stability of $(\mathbf{v}^\mu_\omega, \theta^\mu_\omega, \boldsymbol{\sigma}^\mu_\omega)$ is then deduced from a result of Prüß. (1984).

4. Viscous Heating in a Viscoelastic Fluid

In this section we examine the effect of temperature dependence of the viscosity and of the elasticity parameters.

We assume that the density ρ is independent of temperature, and that the viscosity or the relaxation parameter, denoted by the generic letter χ, obeys an Arrhenius law of the form

$$\chi = \chi_0 \exp\left(\frac{E}{R}\left(\frac{1}{T} - \frac{1}{T_0}\right)\right), \tag{17}$$

or in a simplified form,

$$\chi = \chi_0 \exp\left(-\beta\left(\frac{T}{T_0} - 1\right)\right), \tag{18}$$

where $\beta > 0$ is an empirical parameter. (See for example Bird *et al.*, 1977.)

We introduce the following non-dimensional parameters: the Reynolds number $\mathrm{Re} = \frac{\rho U L}{\eta_0}$ which measures the convection effects relative to the viscous dissipation; the Weissenberg number $\mathrm{We} = \frac{\lambda_0 U}{L}$ which measures the elasticity of the fluid; the Péclet number $\mathrm{Pe} = \frac{\rho C_p U L}{k}$ which measures the bulk transport of thermal energy by convection in the flow direction compared to that by conduction; and the Brinkman number $\mathrm{Br} = \frac{\eta_0 U^2}{k T_0}$ which measures the thermal effects due to viscous heating compared to that due to conduction. U and L are typical velocity and length of the flow. The heat capacity C_p, and the thermal conductivity k are supposed constant, independent of temperature. η_0 and λ_0 denote the total viscosity and the relaxation time of the fluid, both at temperature T_0. We assume that the viscosities and the relaxation time are given by the relations

$$\eta_s(\theta) = \eta_0(1-\omega) f_s(\theta) \qquad \text{(solvent viscosity)},$$

$$\eta_p(\theta) = \eta_0 \omega f_p(\theta) \qquad \text{(polymer viscosity)},$$

$$\lambda(\theta) = \lambda_0 f_\lambda(\theta) \qquad \text{(relaxation time)},$$

in terms of the non-dimensional temperature $\theta = \frac{T - T_0}{T_0}$. The functions f_s, f_p, and f_λ are defined on $[0, +\infty)$, and, if not constant and equal to 1, satisfy

$$f(0) = 1, f > 0, f' < 0, f'' > 0, \lim_{\theta \to +\infty} f(\theta) = 0.$$

Equations for the non-dimensional variables in a bounded domain Ω read as follows:

$$\mathrm{Re}\,\frac{d\mathbf{u}}{dt} + \nabla p = \mathbf{div}\Big((1-\omega) f_s(\theta)\mathbf{D} + \boldsymbol{\tau}\Big), \tag{19}$$

$$\mathrm{div}\,\mathbf{u} = 0, \tag{20}$$

$$\mathrm{We} f_\lambda(\theta) \frac{\mathcal{D}_a \boldsymbol{\tau}}{\mathcal{D}t} + \boldsymbol{\tau} = 2\omega f_p(\theta)\mathbf{D}, \tag{21}$$

$$\mathrm{Pe}\,\frac{d\theta}{dt} = \Delta\theta + \mathrm{Br}\Big[2(1-\omega) f_s(\theta)\mathbf{D} + \boldsymbol{\tau}\Big] : \mathbf{D}, \tag{22}$$

together with boundary conditions and initial conditions for $\mathbf{u}$, $\boldsymbol{\tau}$, and θ. Homogeneous boundary conditions are assumed for $\mathbf{u}$, *i.e.*,

$$\mathbf{u} = \mathbf{0} \quad \text{on} \quad \partial\Omega, \tag{23}$$

and either fixed temperature on the walls, or adiabatic conditions, *i.e.*,

$$\text{either} \qquad \theta = \overline{\theta}_0 \quad \text{on} \quad \partial\Omega, \qquad \text{or} \qquad \frac{\partial\theta}{\partial n} = 0 \quad \text{on} \quad \partial\Omega, \tag{24}$$

where $\frac{\partial}{\partial n}$ denotes the derivative along the outwards pointing unit normal vector to $\partial\Omega$.

4.1. Three-dimensional Flows with Constant Polymer Viscosity and Constant Relaxation Time

Theorem 4.1 (Existence of a local solution)

For all initial conditions $\mathbf{u}_0 \in D(A)$, $\boldsymbol{\tau}_0 \in \mathbf{H}^2_{\mathrm{sym}}(\Omega)$, *and* $\theta_0 \in H^2(\Omega)$, *there exists a time* T^*, *and a solution* $(\mathbf{u}, \boldsymbol{\tau}, \theta)$ *to Problem (19)–(24) satisfying*

$$\mathbf{u} \in L^\infty(0,T^*; D(A)) \cap L^2(0,T^*; \mathbf{H}^3(\Omega)), \tag{25}$$

$$\boldsymbol{\tau} \in L^\infty(0,T^*; \mathbf{H}^2_{\mathrm{sym}}(\Omega)), \tag{26}$$

$$\theta \in L^\infty(0,T^*; H^2(\Omega)) \cap L^2(0,T^*; H^3(\Omega)). \tag{27}$$

The solution $(\mathbf{u}, \boldsymbol{\tau}, \theta)$ *is unique in the class (25)–(27).*
Moreover p *is defined up to an additive constant,*

$$p \in L^\infty(0,T^*; H^2(\Omega)).$$

The proof of the existence of solutions relies on the Schauder fixed point theorem, and on some estimates of the solutions to the linearized equations associated to (19)–(22).

The uniqueness result is deduced from an energy type inequality for the function $\Phi(t) = \mathrm{Re}\,|\mathbf{u}(t)|^2 + \frac{\mathrm{We}}{2\omega}|\boldsymbol{\tau}(t)|^2 + \mathrm{Pe}|\theta(t)|^2$, of the form

$$\Phi'(t) \le F(t)\Phi(t),$$

with $F(t) = C\|\mathbf{u}(t)\|_{\mathbf{L}^\infty(\Omega)}(1 + \|\mathbf{u}(t)\|_{\mathbf{L}^\infty(\Omega)}) \in L^1(0,T^*)$, and C some constant depending on the material parameters, and on the domain Ω of the flow. (See Guillopé *et al.*, 1990a, for similar calculations.)

4.2. One-dimensional Flows with Constant Polymer Viscosity and Constant Relaxation Time

We consider here Couette flows between parallel plates maintained at constant temperatures. The flow has the form

$$\mathbf{u}(\mathbf{x},t) = (0, v(x,t), 0),$$

where $\mathbf{x} = (x,y,z) \in (0,1) \times \mathbf{R} \times \mathbf{R}$. The boundary conditions on the plates are

$$v(0,\cdot) = 0, \qquad v(1,\cdot) = 1.$$

System (19)–(24) reduces to a system of four partial differential equations for the component v of the velocity, for the first difference γ of normal stresses, for the shear stress τ, and for the temperature θ, all defined in $\Omega = (0,1)$.

We first consider the case where there is no heat diffusion. We also assume that the functions f_s, f_p, and f_λ depend on the energy e. The system in terms of v, γ, τ, and e reads as follows:

$$\mathrm{Re}\, v_t - (1-\omega)\Big(f_s(e)v_x\Big)_x = \tau_x, \tag{28}$$

$$\mathrm{We} f_\lambda(e)\gamma_t + \gamma = \mathrm{We} f_\lambda(e)(1-a^2)\tau v_x, \tag{29}$$

$$\mathrm{We} f_\lambda(e)\tau_t + \tau = \Big(\omega f_p(e)v_x - \mathrm{We} f_\lambda(e)\gamma\Big)v_x, \tag{30}$$

$$\mathrm{Re}\, e_t - (1-\omega)f_s(e)\big(v_x\big)^2 = \tau v_x, \tag{31}$$

with boundary conditions

$$v(0,\cdot)=0, \qquad v(1,\cdot)=1, \tag{32}$$

and initial conditions

$$v(\cdot,0)=v_0, \quad \gamma(\cdot,0)=\gamma_0, \quad \tau(\cdot,0)=\tau_0, \quad e(\cdot,0)=e_0. \tag{33}$$

When the functions f_p and f_λ do not depend on energy, *i.e.*, $f_p(e) = f_\lambda(e) = 1$ for all e, then a "steady" shear flow solution of (28)–(33) is given by

$$\overline{v}(x) = x, \qquad \overline{\tau}(x) = \frac{\omega}{1+\kappa^2}, \qquad \overline{\gamma}(x) = \frac{\omega\kappa(1-a^2)^{1/2}}{1+\kappa^2}, \tag{34}$$

with $\kappa^2 = \mathrm{We}^2(1-a^2)$; the corresponding energy $\overline{e} = .\overline{e}(t)$ is given by the relation

$$\int_{e_0}^{\overline{e}(t)} \frac{\mathrm{Re}\; d\xi}{(1-\omega)f_s(\xi)+\overline{\tau}} = t, \quad t \geq 0. \tag{35}$$

We look for solutions to (28)–(33) which are close to the "steady" state (34)–(35).

We also consider the case with heat diffusion and where both plates are maintained at the same temperature. The system then reads as follows:

$$\mathrm{Re}\, v_t - (1-\omega)\Big(f_s(\theta)v_x\Big)_x = \tau_x, \tag{36}$$

$$\mathrm{We} f_\lambda(\theta)\gamma_t + \gamma = \mathrm{We} f_\lambda(\theta)(1-a^2)\tau v_x, \tag{37}$$

$$\mathrm{We} f_\lambda(\theta)\tau_t + \tau = \Big(\omega f_p(\theta)v_x - \mathrm{We} f_\lambda(\theta)\gamma\Big)v_x, \tag{38}$$

$$\mathrm{Pe}\,\theta_t - \theta_{xx} - \mathrm{Br}\,(1-\omega)f_s(\theta)\big(v_x\big)^2 = \mathrm{Br}\,\tau v_x, \tag{39}$$

with boundary conditions

$$v(0,\cdot)=0, \qquad v(1,\cdot)=1, \tag{40}$$

$$\theta(0,\cdot)=0, \qquad \theta(1,\cdot)=0, \tag{41}$$

and initial conditions

$$v(\cdot,0)=v_0, \quad \gamma(\cdot,0)=\gamma_0, \quad \tau(\cdot,0)=\tau_0, \quad \theta(\cdot,0)=\theta_0. \tag{42}$$

When the functions f_p and f_λ do not depend on temperature, *i.e.*, $f_p(\theta) = f_\lambda(\theta) = 1$ for all θ, then steady shear solutions are defined by the following equations:

$$(1-\omega)f_s(\overline{\theta}(x))K(x) + \frac{\omega K(x)}{1+\kappa^2 K(x)^2} = \sigma_0 \quad \text{constant}, \tag{43}$$

$$-\overline{\theta}''(x) = \mathrm{Br}\,\sigma_0 K(x), \tag{44}$$

$$\overline{\theta}(0) = 0, \quad \overline{\theta}(1) = 0, \tag{45}$$

$$\int_0^1 K(x)\,dx = 1, \tag{46}$$

and

$$\overline{v}(x) = \int_0^x K(\xi)\,d\xi, \quad \overline{\tau}(x) = \frac{\omega K(x)}{1+\kappa^2 K(x)^2}, \quad \overline{\gamma}(x) = \frac{\omega.\kappa(1-a^2)^{1/2} K(x)}{1+\kappa^2 K(x)^2}. \tag{47}$$

Note that the constant σ_0 in equations (43) and (44) is unknown, and is determined by conditions (45) and (46). Moreover, σ_0 being fixed, equation (43) is a degree three polynomial function in K, so that there could be one or three solutions depending on the data: K is not *a priori* uniquely defined. (See Guillopé *et al.*, 1990b, Nohel *et al.*, 1990, and Malkus *et al.*, 1994, for similar facts about non-uniqueness).

Here also, we look for solutions of (36)–(42) close to a steady state assumed to be uniquely determined by (43)–(47).

We remark that the stresses are bounded in these one-dimensional problems, with or without heat diffusion.

Lemma 4.1 *Assume that the functions f_p and f_λ are constant, i.e., $f_p(e) = f_\lambda(e) = 1$ for all e.*

Then the solutions to Problem (28)–(33) or to Problem (36)–(42) have bounded stresses, that is

$$\tau, \gamma \in L^\infty((0,1)\times \mathbf{R}_+).$$

The following estimate for τ and γ is easily obtained

$$\begin{aligned}&(1-a^2)\big(\tau(x,t)\big)^2 + \big(\gamma(x,t) - \frac{\omega}{\mathrm{We}}\big)^2 \\ &\le \Big[(1-a^2)\big(\tau(x,0)\big)^2 + \big(\gamma(x,0) - \frac{\omega}{\mathrm{We}}\big)^2\Big]e^{-t/\mathrm{We}} + \big(\frac{\omega}{\mathrm{We}}\big)^2(1-e^{-t/\mathrm{We}}),\end{aligned}$$

for all $(x,t) \in (0,1)\times\mathbf{R}_+$. (See, for example, Renardy *et al.*, 1987, and Guillopé *et al.*, 1990b.)

Remark 4.1 Assume that $f_p(\theta) = f_\lambda(\theta) = 1$, and that $f_s(\theta) = e^{-\beta_s\theta}$ for all θ. Then Problem (36)–(42) admits a unique Newtonian steady state defined by equations (43)–(47) where $\omega = \kappa = 0$. This steady state is given by

$$\overline{v}(x) = \frac{1}{2}\Big(1 + \frac{\tanh(b(x-1/2))}{\tanh(b/2)}\Big), \qquad \overline{\theta}(x) = -\frac{2}{\beta_s}\ln\frac{\cosh(b(x-1/2))}{\cosh(b/2)},$$

where

$$b = 2\sinh^{-1}\Big(\frac{\mathrm{Br}\,\beta_s}{8}\Big)^{1/2}.$$

Linear stability of that state has been studied by Sukanek *et al.*, 1973, and by Ho *et al.*, 1977. The latter concluded that "the low Reynolds number instabilities encountered in polymer processing are not a consequence of thermal effects arising from viscous heating". But their analysis has been done only for Newtonian liquids.

Remark 4.2 Similar problems to Problem (28)–(33) and to Problem (36)–(42) have been studied in the Newtonian case by Dafermos (1982), and Dafermos *et al.* (1982), (1983). In particular some global existence results of unsteady solutions are proven under suitable hypotheses on the function f_s.

5. Concluding Remarks

In this paper, we have focussed on some difficulties inherent to the mathematical theory of flows of viscoelastic fluids. In particular these difficulties are to a large extent independent of thermal effects. In our first example, the Bénard problem, we have used energy type methods to show the (Liapunov) stability of the pure conductive state—at least when the Newtonian viscosity is large enough. Other than in very particular cases where we linearize about the zero solution, or where the problems are one-dimensional (Couette or Poiseuille flows), nonlinear stability of steady states is an open problem, contrary to the case of Newtonian fluid flows: it is kown for Newtonian flows that linear stability of a steady state is governed by the spectrum of the linearized Stokes operator about that state, and that if the flow is linearly stable, then it is Liapunov stable. (See, *e.g.*, Sattinger, 1973.)

As for linear stability of viscoelastic flows, we showed that in the case of weakly viscoelastic fluids (*i.e.*, ω small) it is governed by the spectrum of the linearized operator—this is a C^0 semi-group. Let us point out that M. Renardy (1993) showed that it also holds true in the case of isothermal Couette flows, *i.e.* the linear stability of the isothermal Couette flow is governed by the spectrum of the linearized operator around this steady state.

Numerous numerical studies of the linear stability of flows of Oldroyd type fluids, in particular for Bénard problem, have been done. (See Kolkka *et al.*, 1987, for a review.)

A numerical study of the bifurcation of solutions of Bénard problem has been performed by M. Renardy and Y. Renardy (1992): they show that the problem can be reduced to a finite dimensional one, by applying the center manifold theory for quasilinear hyperbolic partial differential equations.

For flows where viscous heating is important, we have shown the existence of a unique unsteady solution on a small time interval. We also have presented different probems related to one-dimensional shear flows. In the particular cases where the material parameters do not depend on temperature, we have remarked that the stresses are bounded uniformly in time and in space. The existence of global unsteady solutions stays an open problem for viscoelastic fluids with viscous heating.

References

Bazin, V.: 1990, *Contribution à l'analyse numérique d'écoulements de fluides viscoélastiques de type Maxwell dans un canal ondulé*, Thèse de l'Université Paris-Sud, Centre d'Orsay.

Bird, R. B., Armstrong, R. C. and Hassager, O.: 1977, *Dynamics of Polymeric Liquids*, Vol. I: *Fluid Mechanics*, John Wiley & Sons, New York.

Dafermos, C. M.: 1982, 'Global smooth solutions to the initial-boundary value problem for the equations of one-dimensional nonlinear thermoelasticity', *SIAM J. Math. Anal.* Vol no. **13**, pp. 397–408.

Dafermos, C. M. and Hsiao, L.: 1982, 'Global smooth thermomechanical processes in one-dimensional nonlinear thermoviscoelasticity', *Nonlin. Anal., Th. Meth. Appl.* Vol no. **6**, pp. 435–454.

Dafermos, C. M. and Hsiao, L.: 1983, 'Adabiatic shearing of incompressible fluids with temperature-dependent viscosity', *Quart. Appl. Math.* Vol no. **41**, pp. 45–58.

Guillopé, C. and Saut, J.-C.: 1990a, 'Existence results for the flow of viscoelastic fluids with a differential constitutive law', *Nonlin. Anal., Th. Meth. Appl.* Vol no. **15**, pp. 849–869.

Guillopé, C. and Saut, J.-C.: 1990b, 'Global existence and one-dimensional nonlinear stability of shearing motions of viscoelastic fluids of Oldroyd type', *Mod. Math. Anal. Num.* Vol. no. **24**, pp. 369–401.

Guillopé, C. and Saut, J.-C.: 1991, 'Existence and stability of steady flows of weakly viscoelastic fluids', *Proc. Roy. Soc. Edinburgh* Vol. no. **119A**, pp. 137–158.

Ho, T. C., Denn, M. M. and Anshus, B. E.: 1977, 'Stability at low Reynolds number flow with viscous heating', *Rheol. Acta* Vol no. **16**, pp. 61–68.

Iooss, G.: 1971, 'Théorie non linéaire de la stabilité des écoulements laminaires dans le cas de l'"échange des stabilités"', *Arch. Rat. Mech. Anal.* Vol. no. **40**, pp. 166–208.

Iudovich, V. I.: 1966, 'On the origin of convection', *Prikl Mat. Meh.* Vol. no. **30**, pp. 1101–1005, = *J. Appl. Math. Mech.* Vol. no. **30**, pp. 1193–1199.

Iudovich, V. I.: 1967, 'Stability of convection flows', *Prikl Mat. Meh.* Vol. no. **31**, pp. 272–281, = *J. Appl. Math. Mech.* Vol. no. **31**, pp. 294–303.

Joseph, D. D.: 1976, *Stability of Fluid Motions*, Vol. I and II, Springer-Verlag, Berlin.

Kirchgässner, K.: 1975, 'Bifurcation in nonlinear hydrodynamic stability', *SIAM Review* Vol. no. **17**, pp. 652–683.

Kolkka, R. W. and Ierley, G. R.: 1987, 'On the convected linear stability of a viscoelastic Oldroyd B fluid heated from below', *J. Non-Newtonian Fluid Mech.* Vol no. **25**, pp. 209–237.

Larson, R.G., Shaqfeh E. S. G. and Muller, S. J.: 1990, 'A purely elastic instability in Taylor-Couette flow', *J. Fluid Mech.* Vol. no. **218**, pp. 573–611.

Malkus, D. S., Nohel, J. A. and Plohr B., J.: 1994, 'Oscillations in piston-driven shear flow of a non-Newtonian fluid' *in* this volume, pp. 57–71.

Nohel J. A., Pego R. L. and Tzavaras, A. E.: 1990, 'Stability of discontinuous steady states in shearing motions of a non-Newtonian fluid', *Proc. Roy. Soc. Edinburgh* Vol. no. **115A**, pp. 39–59.

Prodi, G.: 1962, 'Teoremi di tipo locale per il sistema di Navier-Stokes e stabilità delle soluzioni stazionarie, *Rend. Sem. Mat. Univ. Padova* Vol. no. **32**, pp. 374–397.

Prüß, J.: 1984, 'On the spectrum of C^0 semigroups', *Trans. Amer. Math. Soc.* Vol. no. **284**, pp. 847–857.

Renardy M.: 1987, 'Existence of slow steady flows of viscoelastic fluids with differential constitutive equations', *Z. Angew. Math. Mech.* Vol. no. **65**, pp. 449–451.

Renardy M.: 1993, 'On the stability of parallel shear flow of an Oldroyd B fluid', *Diff. Integr. Eq.* Vol no. **6**, pp. 481–489.

Renardy M., Hrusa, W. J. and Nohel J. A.: 1987, *Mathematical Problems in Viscoelasticity*, Pitman Monographs and Surveys in Pure and Applied Mathematics Vol no. **35**, Longman Scientific & Technical, Harlow.

Renardy M. and Renardy Y.: 1992, 'Pattern selection in the Bénard problem for a viscoelastic fluid', *Z. Angew. Math. Phys.* Vol. no. **43**, pp. 154–180.

Sattinger D.: 1973, *Topics in stability and bifurcation theory*, Lecture Notes in Mathematics Vol no. **309**, Springer-Verlag, Berlin.

Sukanek, P. C., Goldstein, C. A. and Laurence, R. L.: 1973, 'The stability of plane Couette flow with viscous heating', *J. Fluid Mech.* Vol no. **57**, pp. 651–670.

NUMERICAL SIMULATION AND EXPERIMENTAL RESULTS ON THERMO-ELASTO-HYDRODYNAMIC TILTING-PAD JOURNAL BEARINGS

M. FILLON and J. FRÊNE
University of Poitiers
Laboratory of Solid Mechanics
URA CNRS
40, Av. du recteur Pineau
86022 Poitiers Cedex - FRANCE

Abstract. In that study the two dimensional solution of thermo-elasto-hydrodynamic bearings is obtained by solving simultaneously both the generalized Reynolds equation and the energy equation in the film, and the heat transfer equation in the material. The thermoelastic displacements of all bearing elements are also calculated. The four-shoe tilting-pad journal bearing is studied theoretically and the results are compared with the experimental data. The thermal effects on pressure, film thickness and friction torque are presented. Results show that the thermoelastic displacements of all bearing elements have to be taken into account in order to define the operating bearing clearance and the distortion of internal pad surface. Depending on bearing materials, bearing element temperatures, initial preload ratio and other factors like method of lubricant feeding, the minimum film thickness can decrease or increase when the rotational speed increases. For high preload bearing case, the bearing clearance is small and shape defects could produce great modifications in the bearing behavior.

1. Introduction

Thermal effects are known to play a very important role in lubrication, friction and wear. Excellent bibliographies (Pinkus 1980, 1990, Khonsari 1986 and Fillon et al 1987), have been presented during the last 15 years, so that the introduction is only concerned with the principal studies dealing with thermal effects.

It is commonly accepted that the preliminary studies on thermal effects in lubrication were undertaken in the forties. In fact, thermal effects were analyzed much earlier by Hirn (1854), Beauchamp Tower (1883,1885), Reynolds (1886) and Petrov (1900). These studies show that at constant temperature, the frictional torque is directly proportional to rotational speed and, when the temperature increases, the fluid viscosity decreases and the frictional torque decreases.

More recently Kingsbury (1933) studied both theoretically and experimentally the heat effects in lubricating film. In the theoretical part, Kingsbury proposed to take into account the temperature and the viscosity variations across the film. He showed, by a graphical integration method, that the friction torque and the load carrying capacity are decreased by 60 percent when thermal effects are included.

Ten years later Hagg (1944) solved analytically the problem of Kingsbury and pointed out the importance of the temperature gradient across the film near the moving surface.

The load carrying capacity of parallel surface thrust bearings which cannot be explained by isothermal theory has stimulated a lot of research on thermohydrodynamic effects. The lubricant property variations with temperature is not sufficient to generate the load carrying capacity which was observed experimentally. It is proved today that thermal deformations of material and

J.F. Dijksman and G.D.C. Kuiken (eds.),
IUTAM Symposium on Numerical Simulation of Non-isothermal Flow of Viscoelastic Liquids, 85-99.

particularly these of the fixed pads explain this phenomenon. This behavior was suggested by several authors such as Cameron (1958) and Dowson and Hudson (1963).
More and more sophisticated theoretical models were proposed. The simple one was to use an effective temperature and the corresponding effective viscosity through an evaluation of the dissipated power using an isothermal theory. Then, several authors proposed the use of a particular fluid film energy assuming adiabatic conditions and constant viscosity across the film.
A very important step was achieved in 1962, when Dowson (1962) proposed a generalized Reynolds equation which permits to take into account simultaneously the viscosity variations and the density variations of the lubricant both along and across the film. This new equation when associated with both the energy equation and the heat transfer equation along with the thermal boundary conditions adequately describes the thermal phenomena in hydrodynamic lubrication. The film modelization could be either of two or three dimensional form and by using more and more efficient computers, any lubricant film could be studied accurately.
Contemporary, theoretical studies have to modelize the film thickness accurately and take into account the thermal elastic displacements of all the bearing elements. The influence of the lubricant feeding method on bearing performances have been studied recently by Fillon et al (1993) and Tanaka (1993): under severe operating conditions, the directed lubrication permits to reduce both the operating bearing clearance and the maximum babbitt temperature when compared to the pressurized oil system.
In this work, the theory is presented for the tilting-pad journal bearing case lubricated by flooding the bearing (pressurized oil method). The thermoelastic bearing element displacements and the reversed flow at the inlet are taken into account. The theoretical results are then compared to experimental data.

2. Theoretical analysis

In this study, thermal and deformation problems are solved in the mid-plane bearing, assuming that the axial variation of both temperature and solid displacements are very small and could be neglected.

2.1 Basic Equations

Under classical assumptions, the generalized Reynolds equation in non dimensional form can be written (Boncompain et al, 1987) :

$$\frac{\partial}{\partial\theta}\left(\bar{h}^3\bar{G}\frac{\partial\bar{p}}{\partial\theta}\right)+\left(\frac{R_a}{L}\right)^2\frac{\partial}{\partial\bar{z}}\left(\bar{h}^3\bar{G}\frac{\partial\bar{p}}{\partial\bar{z}}\right)=\frac{\partial}{\partial\theta}\left(\bar{h}\left(1-\frac{\bar{I}_2}{\bar{J}_2}\right)\right) \tag{1}$$

where the non dimensional functions $\bar{I}_2$, $\bar{J}_2$ and $\bar{G}$ are defined by :

$$\bar{I}_2=\int_0^1\frac{\bar{y}}{\bar{\mu}}d\bar{y},\qquad \bar{J}_2=\int_0^1\frac{d\bar{y}}{\bar{\mu}},\qquad \bar{G}=\int_0^1\frac{\bar{y}}{\bar{\mu}}\left(\bar{y}-\frac{\bar{I}_2}{\bar{J}_2}\right)d\bar{y} \tag{2}$$

θ, $\bar{y}$ and $\bar{z}$ are the non dimensional co-ordinates (fig. 1), R_a and L are respectively the journal radius and the bearing length, $\bar{\mu}$ and $\bar{p}$ are the non dimensional viscosity and pressure. $\bar{h}$ is the non dimensional film thickness defined by :

$$\bar{h}(\theta)=1+\varepsilon_i\cos\theta+\delta(\theta) \tag{3}$$

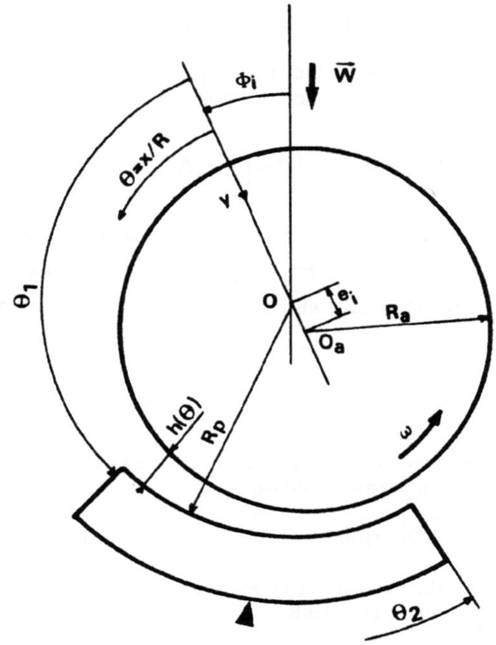

Fig. 1 : Co-ordinate system

where $\varepsilon_i = e_i / C_p$ is the relative eccentricity of the non deformed pad and $\delta(\theta)$ is the non dimensional film thickness variation due to bearing element displacements.

The lubricant viscosity-temperature variation is given by this relationship :

$$\bar{\mu}(\theta, \bar{y}) = \frac{\mu}{\mu_o} = e^{-\beta' T_o (\bar{T} - 1)} \tag{4}$$

Classical form of the two-dimensional energy equation used in the lubricant is :

$$P_e \left(\bar{u} \frac{\partial \bar{T}}{\partial \theta} + \bar{v} \frac{\partial \bar{T}}{\partial \bar{y}} \right) = \frac{1}{\bar{h}^2} \frac{\partial^2 \bar{T}}{\partial \bar{y}^2} + B_r \frac{\bar{\mu}}{\bar{h}^2} \left(\frac{\partial \bar{u}}{\partial \bar{y}} \right)^2 \tag{5}$$

In this equation, P_e is the Peclet number, B_r is the Brinckman number, $\bar{u}$ and $\bar{v}$ are the velocity components of the fluid and $\bar{T}$ is the non dimensional temperature in the film.

The heat transfer equation in the pad is written as follows :

$$\frac{\partial^2 \bar{T}}{\partial \bar{r}^2} + \frac{1}{\bar{r}} \frac{\partial \bar{T}}{\partial \bar{r}} + \frac{1}{\bar{r}^2} \left(\frac{\partial^2 \bar{T}}{\partial \theta^2} \right) = 0 \tag{6}$$

where θ and $\bar{r}$ are the non dimensional co-ordinates used in the pad and $\overline{T}$ is the non dimensional temperature in the pad. The shaft temperature is constant and is either given or determined assuming that value is equal to the average film temperature.

Temperature and pressure fields change the initial geometry of bearing elements. The thermoelastic displacements of the pads are determined using the thermoelasticity law :

$$\varepsilon_{ij} = \frac{1+\nu}{E}\sigma_{ij} - \frac{\nu}{E}(\sigma_{kk}) + \alpha_a \Delta T(M)\delta_{ij} \tag{7}$$

In this relation, ε_{ij} and σ_{ij} are respectively the strain and stress tensors, E and ν are respectively the modulus of elasticity and the Poisson's ratio, α_a the thermal expansion coefficient and $\Delta T(M)$ is the temperature variation at point M, considered with regard to a reference temperature.

2.2. Boundary Conditions

The classical Reynolds boundary conditions are associated with the Reynolds equation.
The boundary conditions on temperature are as follows :

- On the external shaft surface (film shaft interface), the temperature is imposed.

- On the outer parts of the pads, free convection and radiation hypothesis is employed. For example, at the external pad radius, this yields :

$$\left.\frac{\partial \overline{T}}{\partial \bar{r}}\right|_{\bar{r}=\overline{R}_{ep}} = -\frac{H_p R_p}{K_p}\left(\left.\overline{T}\right|_{\bar{r}=\overline{R}_{ep}} - \overline{T}_{bath}\right) \tag{8}$$

where $R_p = R_a + C_p$ and $R_{ep} = R_p + E_p$ are the internal and external pad radius, H_p and K_p are respectively the heat convection coefficient and the pad thermal conductivity and $\overline{T}_{bath}$ is the non dimensional temperature of the oil bath in the chamber between pads.

- On the surface between the film and the pad, the temperature is obtained by the heat flux continuity condition which gives :

$$\left.\frac{\partial \overline{T}}{\partial \bar{r}}\right|_{\bar{r}=1} = -\frac{R_p}{C_p}\frac{K_h}{K_p}\frac{1}{\bar{h}}\left.\frac{\partial \overline{T}}{\partial \bar{y}}\right|_{\bar{y}=0} \tag{9}$$

where C_p is the radial pad clearance and K_h the lubricant thermal conductivity.

- In the inlet zone of the film, the fluid temperature is assumed to vary parabolically across the film between T_p and T_a which are respectively the pad and the shaft temperature. In the chambers between pads, the heat balance permit to take into account the lubricant mixing between the hot oil carried over from the previous pad and the cold oil feeding in the bearing. The coefficient of hot oil carry over which could vary is generally equal to 0.85 for the pressurized oil feeding method.

2.3. Method of Solution and Procedure

The pressure distribution in the film and the temperature field both in the film and in the pads are obtained by iterative techniques. Finite differences and Gauss-Seidel methods with over relaxation are employed to solve the generalized Reynolds equation. Negative pressure terms are set equal to zero when they appear in the iterations needed to satisfy the Reynolds boundary conditions. Finite differences and Gauss-Seidel methods with over relaxation are also used to solve heat transfer equation in the pads.

The energy equation (5) looks like a differential parabolic equation where θ plays the role of the time in the non-stationary problems. An implicit finite difference method is used to solved this equation and the Richtmyer technique (1957) is employed when the coefficient of $\partial^2\overline{T}/\partial\overline{y}^2$ is positive i.e., when the fluid velocity component $\overline{u}$ is positive.

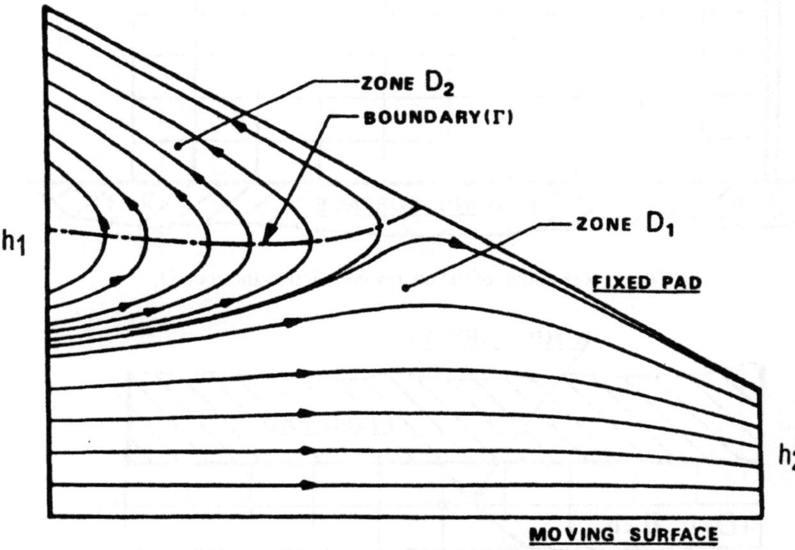

Fig. 2 : Stream lines in thrust bearing for $h_1/h_2 = 4$

When a reversed flow occurs, at large relative pad eccentricities for example, $\overline{u}$ is negative in the inlet zone of the film and classical solutions do not converge. In this case, the fluid film is divided into two zones : these zones D_1 and D_2 are shown on fig. 2 for the case of a thrust bearing operating with a film thickness ratio of the leading edge to the trailing edge equal to 4.
In the first zone which corresponds to the zone where $\overline{u}$ is positive, the energy equation is solved using the iterative Richtmyer technique. The initial temperature values are the boundary conditions at the leading edge of the film and a given value at points situated in D_2 near the boundary (fig. 3).

In the second zone, the same iterative technique is used but the computing process is started from the boundary using the initial temperature at points situated in D_1 near the boundary (fig. 4). The iterative process is performed alternatively in both zones and converges very quickly. Figure 5 shows, as an example, results obtained for a plane thrust bearing. A constant temperature is imposed on the moving surface and heat flux continuity conditions are used on the fixed pad. The full lines are the isothermal lines and the dash lines are the heat flux lines.

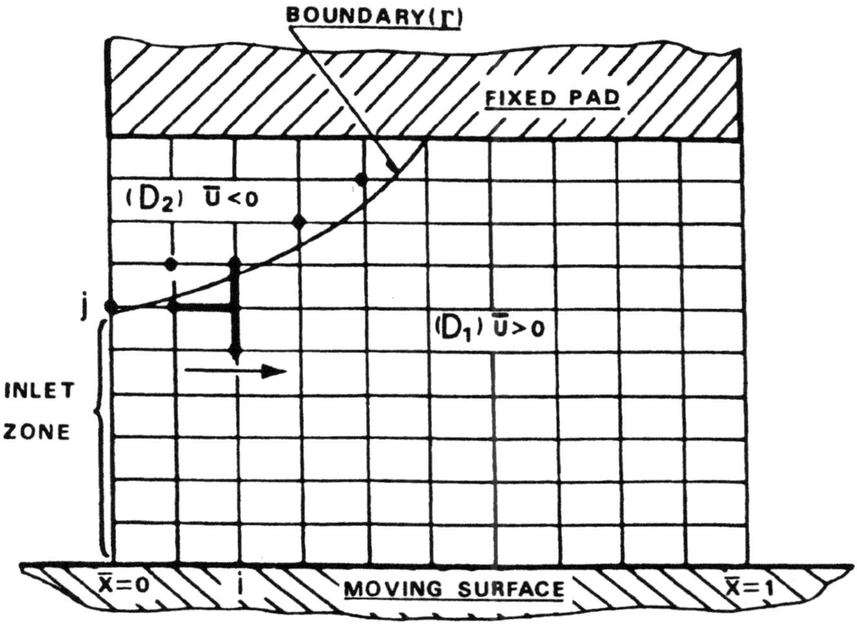

Fig. 3 : Resolution of the energy equation in the zone D_1

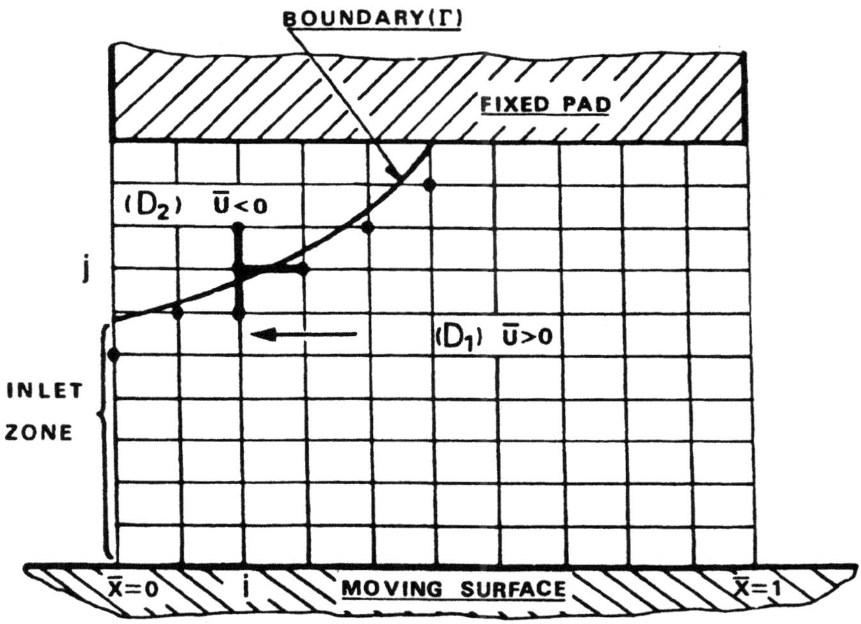

Fig. 4 : Resolution of the energy equation in the zone D_2

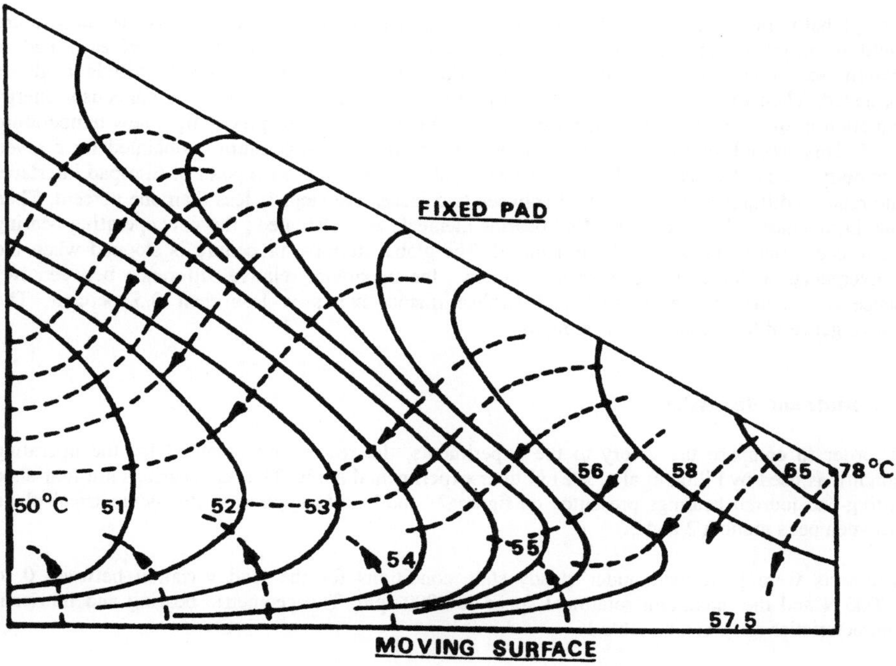

Fig. 5 : Isotherms and heat flux lines in the film for $h_1/h_2 = 4$

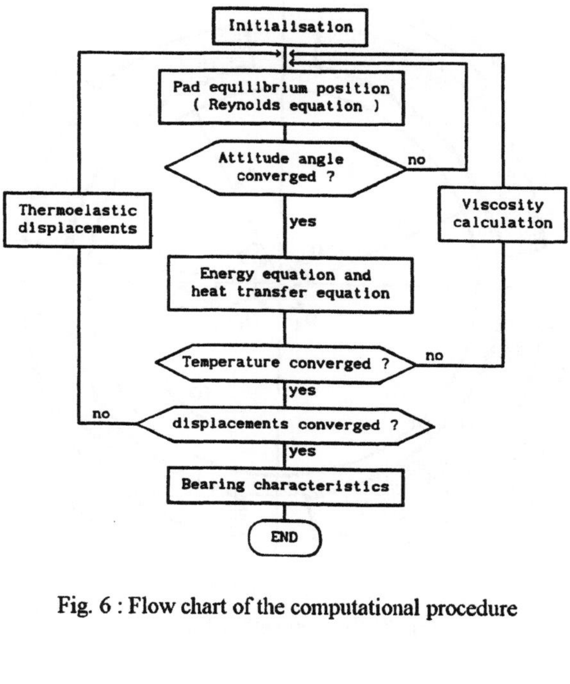

Fig. 6 : Flow chart of the computational procedure

The global iterative scheme (fig. 6) is as follows : for an initial global eccentricity the temperature field is initialized to a given temperature. successively, equilibrium position of each pad is determined solving Reynolds equation until the convergence of the global attitude angle is achieved. Then the calculated fluid velocity components permit to solve simultaneously energy equation in the film and heat transfer equation in the pads and thus producing a new temperature field. This second loop is stopped when the convergence on temperature is obtained for a given geometry (calculated bearing element displacements), i.e. when at each point of film-pad interface, the relative difference of temperature between two successive steps is less than one percent. Then the thermoelastic displacements of all bearing elements are calculated ; the new operating bearing clearances relative to each pad are deduced. The global iterative procedure is stopped when the convergence on the displacements is obtained ; the maximum relative difference between two successive steps of the internal pad surface displacements is less than 0.5 percent. The convergence of this process is fairly rapid.

3. Results and discussion

In order to compare the theory to the experiments, the results are obtained for the operating conditions used by Fillon et al (1992) in their experimental study. The test bearings are four-shoe tilting-pad journal bearings presented on figure 7 ; the load is applied in the symmetrical plane between pads number 2 and 3.

The tests were performed under steady state conditions for the load variation between 0 to 10000 N and the maximum rotational speed is 4000 rpm. The geometric bearing and lubricant characteristics are given on table 1.

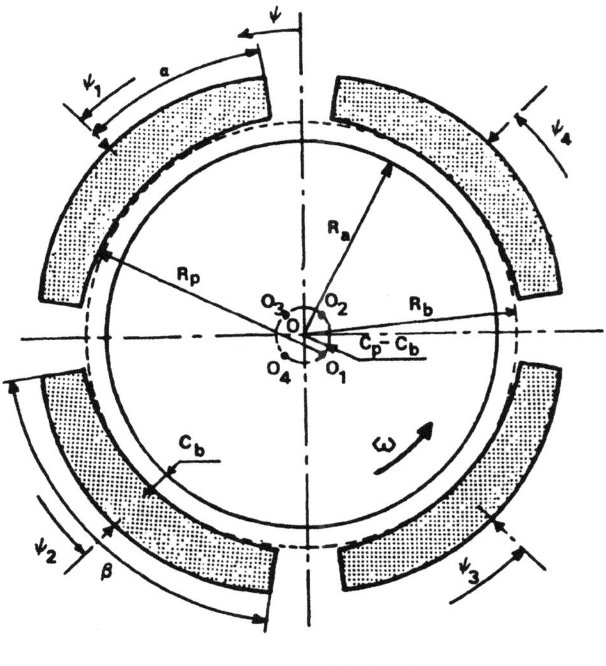

Fig. 7 : Geometry of the bearing

Table 1 : Bearing and lubricant characteristics

		Bearing A	Bearing B
Journal radius	(10^{-3} m)	$R_a = 49.886$	$R_a = 49.840$
Radial pad clearance at 20°C	(10^{-3} m)	$C_p = 0.148$	$C_p = 0.160$
Radial bearing clearance at 20°C	(10^{-3} m)	$C_b = 0.079$	$C_b = 0.079$
Preload ratio at 20°C		$m = 0.47$	$m = 0.47$
Bearing length	(10^{-3} m)	$L = 70.$	
Pad thickness	(10^{-3} m)	$E_p = 20.$	
Angular amplitude of pads	(°)	75.	
Relative pivot position on pads		0.5	
Pivot positions in bearing	(°)	$\Psi 1 = 45.$ $\Psi 2 = 135.$ $\Psi 3 = 225.$ $\Psi 4 = 315.$	
Feeding lubricant temperature	(°C)	$T_o = 40.$	
Feeding lubricant pressure	(MPa)	$P_o = 0.05$	
Lubricant viscosity at 40°C	(Pa.s)	$\mu_o = 0.0277$	
Lubricant density at 40°C	(Kg.m^{-3})	$\rho = 860$	
Temperature viscosity coefficient	(°C^{-1})	$\beta' = 0.034$	
Lubricant thermal conductivity	(W.m^{-1}.°C^{-1})	$K_h = 0.13$	
Lubricant specific heat	(J.kg^{-1}.°C^{-1})	$C_o = 2000$	
Pad thermal conductivity	(W.m^{-1}.°C^{-1})	$K_p = 45.$	
Heat convection coefficient	(W.m^{-2}.°C^{-1})	$H_p = 100.$	
Modulus of elasticity	(MPa)	$E = 210{,}000$	
Poisson's ratio		$\nu = 0.29$	
Thermal expansion coefficient	(°C^{-1})	$\alpha_a = 12.$	
Shaft temp. (N=4000rpm;W=10000N)	(°C)	$T_a = 65.5$	$T_a = 78.0$

The results presented on the following figures are obtained for the most severe operating conditions (W = 10,000N ; N = 4000 rpm). Figures 8 to 12 are concerned with the configuration of bearing A thus figures 13 and 14 present results obtained with the configuration of bearing B.

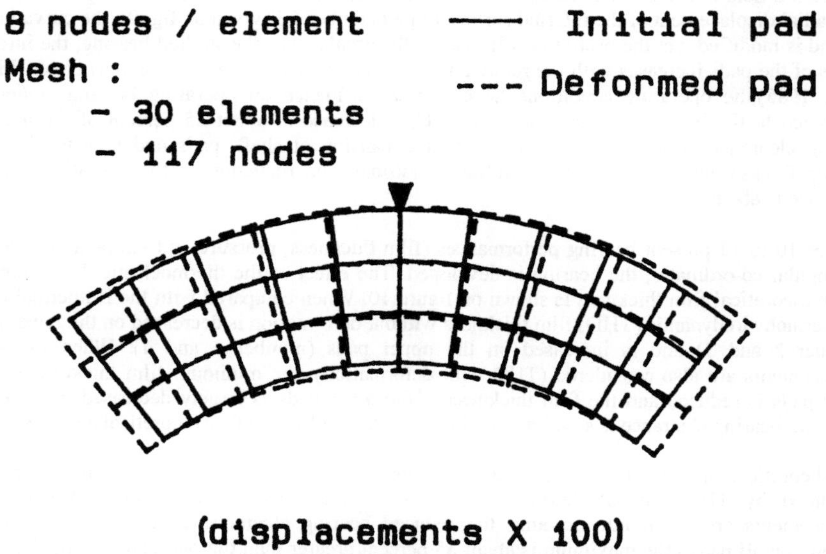

Fig. 8 : Theoretical distortion of pad number 2 (bearing A)

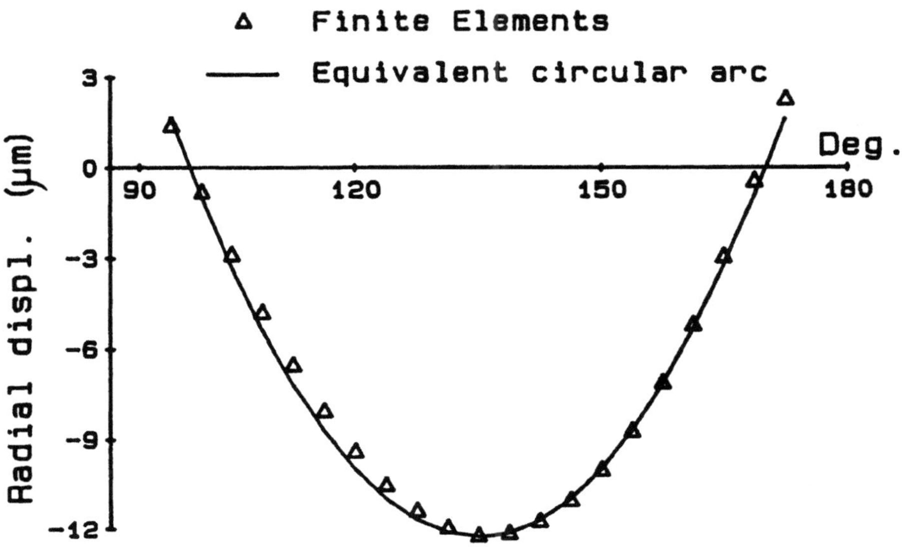

Fig. 9 : Radial displacement of the internal surface of pad number 2 (bearing A)

Figure 8 shows the theoretical pad distortion for the loaded pad number 2. Maximum radial displacement of the internal surface is on the pivot line and reached 12 μm. The displacements due to pressure field are small and are equal to 10 to 20 percent of these due to the temperature field. The radial displacement of the internal surface of pad number 2 is given on fig. 9 ; the curvature of the pad is modified, yet the profile is still essentially circular. For the studied bearing, the internal radius of the pads increases with the rotational speed. As it was shown in a previous paper (Fillon et al, 1985),the operating conditions have a great influence on operating bearing clearance. Compared to the bearing characteristics at 20°C, a decrease of about 25 percent of the average bearing clearance is calculated due to thermal expansion of shaft, pads and housing. For the bearing B case and for the same operating conditions, the reduction of the operating bearing clearance is about 50 percent.

Figures 10 to 12 present bearing performances (film thickness, pressure and temperature) versus the angular co-ordinate ; the bearing is developed. The effect of the thermoelastic displacements on the theoretical film thickness is shown on figure 10. When compared with the isothermal case, the thermohydrodynamic (THD) film thickness without deformation is decreased on the lower pads (number 2 and 3) and is increased on the upper pads (number 1 and 4). When the solid displacements are also considered (THD with deformation), the minimum film thickness on the lower pads is reduced and the film thickness of the other pads is largely decreased because the operating bearing clearance is smaller than the one obtained by the other theoretical analysis.

The theoretical pressure in the mid-plane bearing is shown on figure 11. The pressure field calculated by THD without deformation and isothermal theories are similar. When solid displacements are taken into account, the reduced bearing clearance leads to an increase of pressure on all pads : the maximum is about 35 percent greater than the one obtained by the other solutions.

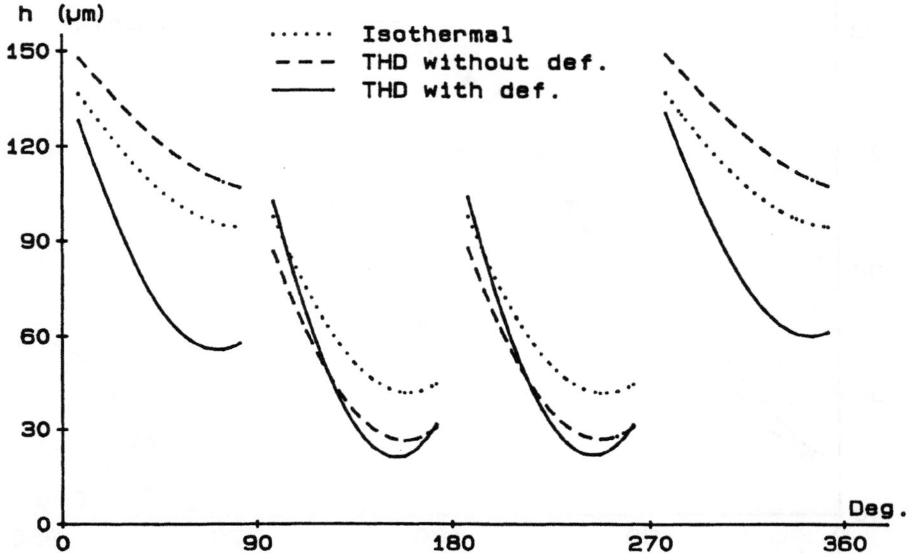

Fig. 10 :Theoretical film thickness (bearing A)

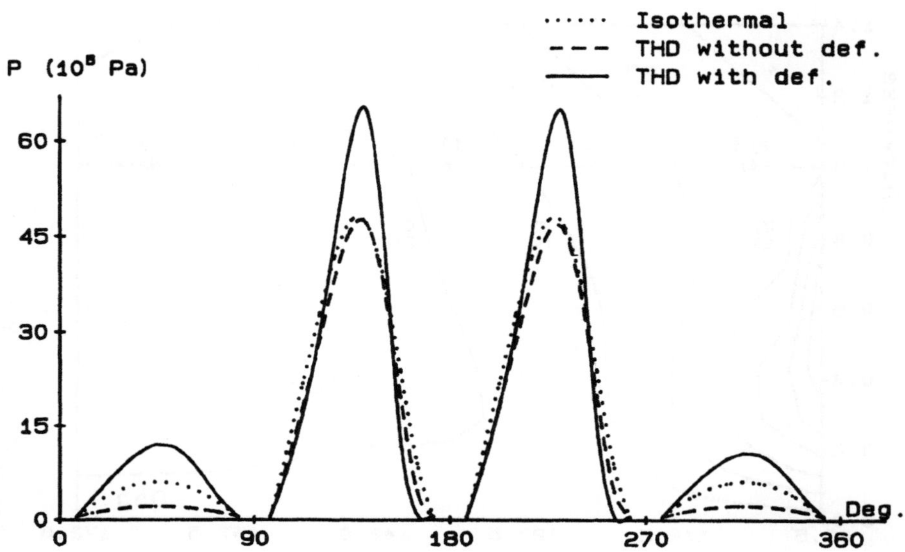

Fig. 11 : Theoretical pressure in mid-plane bearing (bearing A)

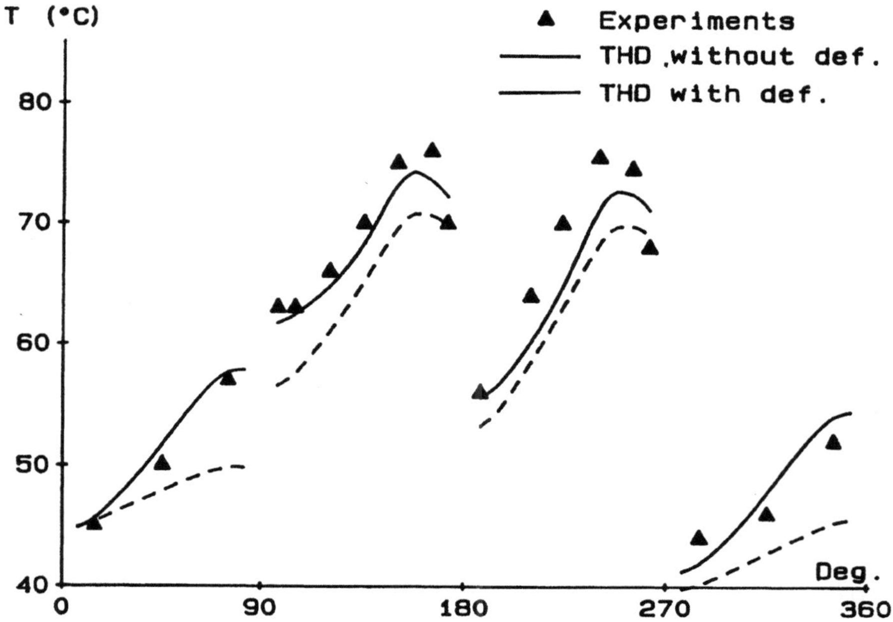

Fig. 12 :Theoretical and experimental babbitt temperatures (bearing A)

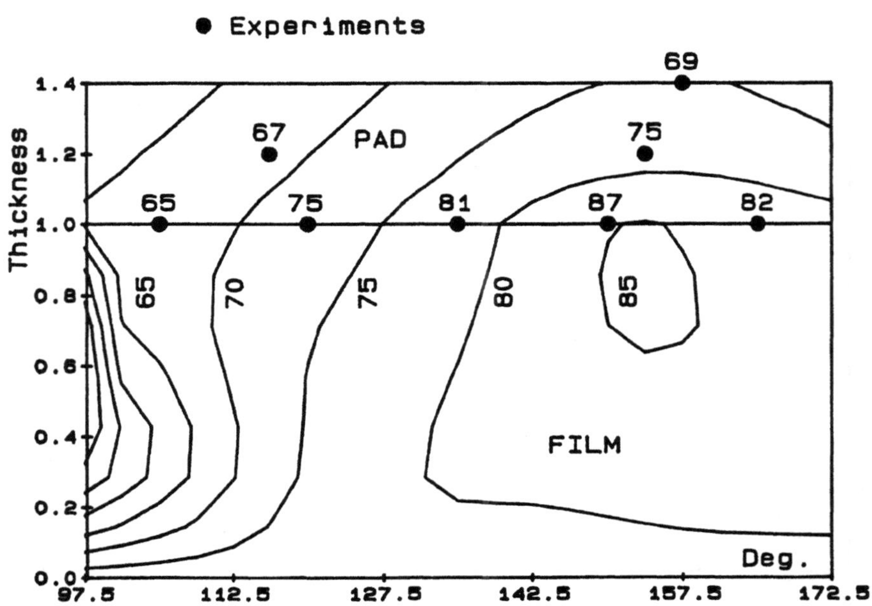

Fig. 13 : Isotherms in the film and in the pad number 2 (bearing B)

Figure 12 presents both theoretical and experimental temperature variations of the internal pad surface. Discrepancies are noted between theoretical results obtained with and without deformation. Theoretical results including thermoelastic displacements of all bearing elements are in good agreement with experimental data. The effect of the reduced bearing clearance on temperature profile is shown on this figure : when solid displacements are not taken into account, the difference between theoretical temperature and experimental data reaches 10°C.
Figure 13 gives the isothermal chart in the film and in the pad number 2 of bearing B, for a rotational speed of 4000 rpm and a load equal to 10,000 N. Different scales are used for the film and for the pad ; because of the variable used, the film thickness appeared to be constant. The experimental temperatures are also noted on this figure. The calculated maximum film temperature is located at 75 percent of the pad length in the circumferential direction, near the internal surface of the pad. The maximum babbitt temperature is measured in this zone. The profile of the isothermal curves in the inlet zone of the film is due to the assumption of parabolic variation across the film thickness. In the inlet zone, the shaft gives heat to the film, thus in the outlet zone the film gives heat to the shaft. The theoretical temperatures are in good agreement with experimental data except for the back surface of the pad due to the assumption of constant temperature of the fluid in the chamber.

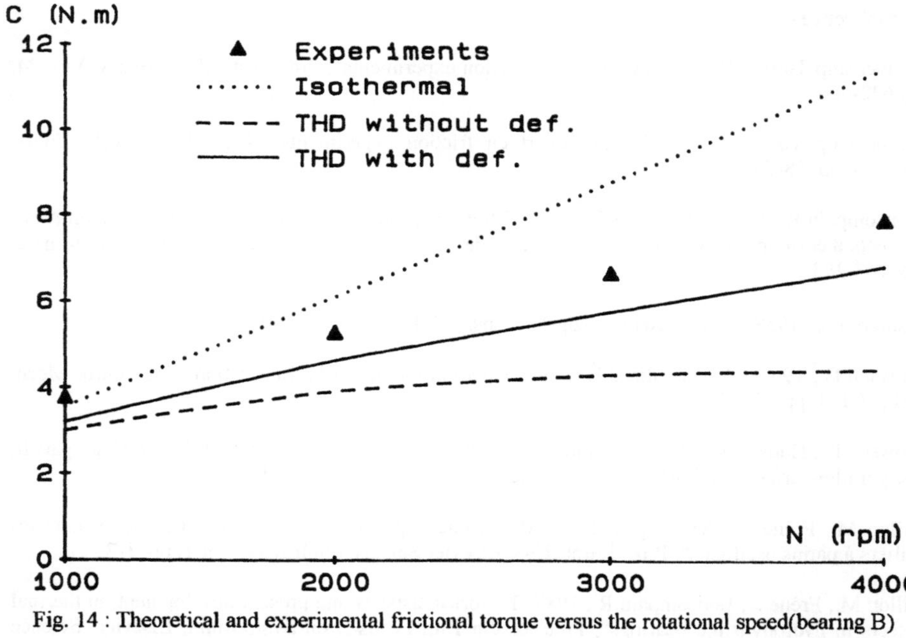

Fig. 14 : Theoretical and experimental frictional torque versus the rotational speed(bearing B)

On figure 14, the frictional torque measured on the bearing B is compared to the ones calculated with three theoretical model : isothermal, thermohydrodynamic with and without deformation. The results are given for the maximum applied load versus the rotational speed. The frictional torque increased with the rotational speed. The isothermal model overestimates the frictional torque whereas the thermohydrodynamic model without deformation underestimates it. The best agreement is obtained with the thermohydrodynamic model with deformation. The little discrepancies between theoretical results and experimental data could be due to the churning oil in the chambers between pads which is not taken into account in the theoretical analysis.

4. Conclusion

The results show that the operating conditions have a great influence on bearing temperature and that the maximum babbit temperature is located in a zone between 70 and 90 percent of circumferential pad length.
The radial operating bearing clearance reduces when both the rotational speed and the applied load increase.
The comparison between theoretical results and experimental data shows that the thermohydrodynamic analysis alone , which permits to include the viscosity variation is not sufficient to predict bearing performances ; the displacements of all bearing elements have to be taken into account.
A future extension of the theoretical model will consist in the generalization to the non laminar flow.

5. Acknowledgements

That work was partially supported by the Electricité de France, Direction des Etudes et Recherches, Chatou, under the contract number P33L31/2F6640.

6. References

Beauchamp Tower, 1883, "First report on friction experiments", Proc. Inst. Mech. Engrs, Vol. 34, pp.632-666.

Beauchamp Tower, 1885, "Second report on friction experiments", Proc. Inst. Mech. Engrs, Vol. 36, pp. 58-70.

Boncompain R., Fillon M., Frêne J., 1987, "Effets thermiques dans les paliers hydrodynamiques - Aspects théoriques et expérimentaux", Journal de Mécanique théorique et appliquée, Vol. 6, n° 2, pp. 253-293.

Cameron A., 1958, "The viscosity wedge", Trans. ASLE, Vol. 1, pp. 248-253.

Dowson D., 1962, "A generalized Reynolds equation for fluid film lubrication", Int. Journ. Mech. Sci., Vol. 4, pp.159-170.

Dowson D., Hudson J., 1963, "Thermohydrodynamic analysis of the infinite slider-bearing : part II, the parallel-surface bearing", Proc. Inst. Mech. Engrs, Vol. 34, pp.45-51.

Fillon M., Frêne J., Boncompain R., 1985, "Etude expérimentale de l'effet thermique dans les paliers à patins oscillants", Proc. Eurotrib'85, Elsevier Science Publishers, Vol.1, pp. 1-7.

Fillon M., Frêne J., Boncompain R., 1987 "Historical aspects and present development on thermal effects in hydrodynamic bearings", Proc. of the 13th Leeds-Lyon Symposium, Elsevier. Science Publishers, Tribology Series 11, pp. 27-47.

Fillon M., Bligoud J.C., Frêne J., 1992, "Experimental study of tilting-pad journal bearings - Comparison with theoretical thermoelastohydrodynamic results", ASME, Journ. of Tribology, Vol. 114, pp. 579-588.

Fillon M., Bligoud J.C., Frêne J., 1993, "Influence of the lubricant feeding method on the thermoelastohydrodynamic characteristics of tilting-pad journal bearings", Proc. of the 6th International Congress on Tribology, Eurotrib'93, Vol. IV, pp. 7-12.

Hagg A.G., 1944, "Heat effects in lubricating films", Journ. of App. Mech., N° 11, pp. 72-76.

Hirn G.A., 1854, "Sur les principaux phénomènes que présentent les frottements médiats, et sur les diverses manières de déterminer la valeur mécanique des matières employées au graissage des machines", Bull. Soc. Ind. de Mulhouse, Tome XXVI, pp. 188-277.

Khonsari M., 1986 a, "A review of thermal effects in hydrodynamic bearings - Part I : slider and thrust bearings", ASLE Trans., Vol. 30, N° 1, pp. 19-25.

Khonsari M., 1986 b, "A review of thermal effects in hydrodynamic bearings - Part II : journal bearings", ASLE Trans., Vol. 30, N° 1, pp. 26-33.

Kingsbury A., 1933, "Heat effects in lubricating films", Mech. Engrs., N° 55, pp. 685-688.

Petrov N., 1900, "Résultats les plus marquants de l'étude théorique sur le frottement médiat", Revue de Mécanique, Vol. VII, pp. 571-602.

Pinkus O.and Wilcock D.J., 1980, "Thermal effectsin fluid bearings", Proc. of the 6th Leeds-Lyon Symposium, Mech. Eng. Publi., pp. 3-23.

Pinkus O., 1990, "Thermal aspects of fluid film tribology", ASME Press New York, 506 p.

Reynolds O., 1886, "On the theory of lubrication and its application to M. Beauchamp Tower's experiments", Phil. Trans. Roy. Soc., Series A 177, pp. 135-217.

Richtmyer R.D., 1957, "Difference methods for initial value problems", Interscience Publishers, Inc. New York, 101 p.

Tanaka M. "The effect of nozzle design and less pad assembly on the thermo-hydrodynamic performance of tilting pad journal bearings with spot lubrication",Proc. of the 6th International Congress on Tribology, Eurotrib'93, Vol. IV, pp. 1-6.

NON-ISOTHERMAL EFFECTS IN CAPILLARY VISCOMETRY

S. GUNTER* and T. N. PHILLIPS
Department of Mathematics
University of Wales
Aberystwyth SY23 3BZ
United Kingdom

Abstract. Viscous dissipation is known to provide a source of error in the determination of viscosity using the conventional method of capillary flow analysis. This is because the flow in the capillary is assumed to be at a constant temperature. The implications of viscous dissipation for viscometry are investigated by numerically simulating the flow in a capillary viscometer using a finite difference discretization of the governing equations of motion. The method of false transients is used to accelerate the scheme to convergence. The simulations confirm that viscous dissipation causes local temperature rises in the fluid particularly near the capillary wall. The viscosity function determined from the standard capillary flow analysis is therefore in error. The inverse problem in which viscosity is determined under non-isothermal conditions is addressed.

Key words: Viscous dissipation, capillary viscometry, Biot condition, inverse problem

1. Introduction

Most rheological processes are widely assumed to take place at a uniform, constant operating temperature, *i.e.* the flow is assumed to be isothermal. In such cases the effects of temperature on the flow and on the rheological properties of the fluid are neglected. This assumption is made, for example, when determining the viscosity of a fluid using a capillary viscometer and capillary flow analysis. However, in practice, viscous dissipation can arise in regions of high shear-rates near the capillary wall. Some of the heat generated is conducted to the surroundings and some remains in the fluid where it is transported by conduction and convection. The amount of heat conducted to the surroundings is dependent on the die wall (its geometric dimensions and thermal conductivity). The transport of heat within the fluid is described by an energy equation. Significant flow re-arrangement can occur since the momentum and energy equations are coupled through the temperature dependence of viscosity. Thus the flow becomes non-isothermal.

In this paper we investigate these effects by numerically simulating the flow through a capillary. In the simulations the set of governing equations also includes the energy equation in which a viscous dissipation term is included. The results show that non-isothermal effects need to be taken into account if the viscosity is to be calculated accurately. A method for obtaining a more accurate viscosity function is outlined. The procedure involves finding the solution to an inverse problem.

We consider the problem of non-isothermal flow coupled with heat conduction

* Present address: Department of Computer Science, University College, Singleton Park, Swansea, SA2 8PP.

J.F. Dijksman and G.D.C. Kuiken (eds.),
IUTAM Symposium on Numerical Simulation of Non-isothermal Flow of Viscoelastic Liquids, 101-132.

in the die wall, which we term the *Full Problem*. We also consider an alternative formulation, the *Biot Problem*, in which we approximate the conduction problem in the die wall by a boundary condition involving the Biot number.

In the remainder of the introduction we give a brief literature review of theoretical and experimental results for non-isothermal capillary flow. Brinkman[1] calculated the developing temperature distribution of flow in a finite capillary of a Newtonian fluid with temperature independent viscosity. Bird[2] extended this work to include temperature independent power-law fluids. Kearsley[5] and Martin[12] have solved the problem of fully developed flow of a fluid with temperature dependent viscosity in an infinitely long capillary. In this case the heat transfer by convection can be neglected: heat conduction balances dissipation. Perturbation analyses by Galili *et al*[19, 20] and Heiber[21] obtained first order approximations to the thermally developing flow of temperature and shear-rate dependent (power-law) viscosity fluids.

In experimental investigations of viscous dissipation in capillary flow Marshall *et al*[7], van Leeuwen[11] and Kim & Collins[13] have attempted to make temperature measurements of the flowing fluid using thermocouples. Their efforts have met with little success, primarily due to the problem of viscous dissipation on the thermocouple itself and disruption of the flow field. A more successful method has been employed by Cox & Makosko[17], who used an infrared pyrometer to measure melt surface temperatures of the extrudate.

The experimental results of Gerrard & Philippoff[6] and the numerical results of Gerrard *et al*[9, 10] and Gee & Lyon[3] all indicate increasing flow-rate as a result of viscous dissipation, for a set inlet pressure.

2. Capillary Viscometry

The capillary viscometer is one of the most popular instruments used by rheologists to determine the viscosity of polymer melts since viscosity data can be obtained up to high shear-rates and temperatures.

The theory of capillary viscometry is based on the assumption of fully-developed, isothermal flow in the capillary, where the pressure gradient along the capillary is regarded as constant. The shear-rate at the capillary wall can then be expressed as [18]

$$\dot{\gamma}_w = \frac{4Q}{\pi R_c^3}\left[\frac{3}{4} + \frac{1}{4}\frac{\mathrm{d}\ln Q}{\mathrm{d}\ln \sigma_w}\right]. \tag{1}$$

Q is the volumetric flow-rate, R_c is the capillary radius and σ_w is the wall shear stress, given by

$$\sigma_w = \frac{R_c}{2}\frac{\mathrm{d}p}{\mathrm{d}L}, \tag{2}$$

where $\frac{\mathrm{d}p}{\mathrm{d}L}$ is the (assumed constant) pressure gradient along the capillary length, L. The term in square brackets in (1) is the Weissenberg-Rabinowitsch correction. It is equal to one for Newtonian fluids and is greater than one in the case of shear thinning fluids.

The fluid viscosity at the wall shear-rate is then given by

$$\eta(\dot{\gamma}_w) = \frac{\sigma_w}{\dot{\gamma}_w}. \tag{3}$$

Using the above results for a range of temperatures, a series of $\eta(\dot{\gamma})$ curves can be determined. Fig. 1 shows an example of this; the $\eta(\dot{\gamma})$ curves for the polycarbonate melt Makrolon, as determined by Walters[27].

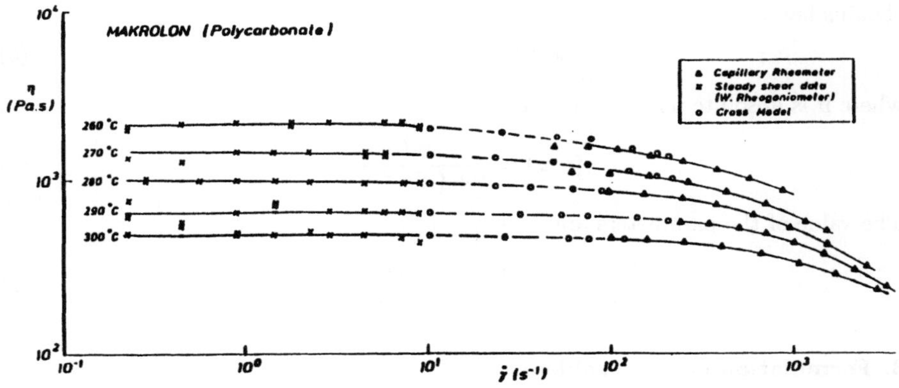

Fig. 1. Variation of viscosity with shear-rate and temperature for the polycarbonate melt Makrolon.

The variation of viscosity with shear-rate and temperature, for this example, can be described by a *master curve*, $\eta(\dot{\gamma}, \vartheta)$, by plotting the data in terms of reduced variables[8]:

$$\eta_R = \frac{\eta}{\eta_0} \quad \text{and} \quad \dot{\gamma}_R = \eta_0 \dot{\gamma} \tag{4}$$

(see Fig. 2), where η_0 is the zero-shear viscosity determined from viscosity measurements at low shear-rates.

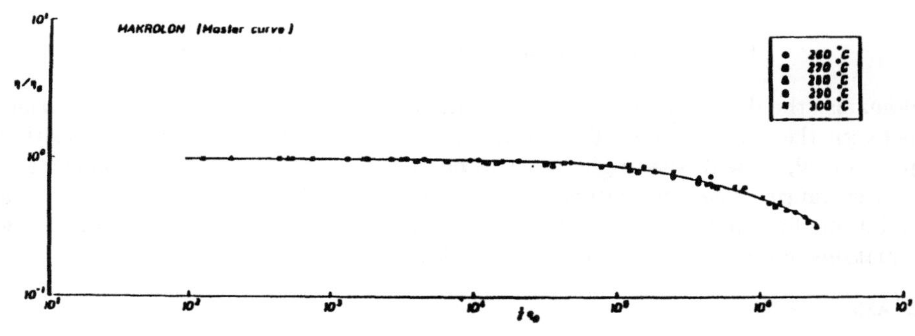

Fig. 2. Makrolon master curve.

The curve of Fig. 2 can be described by the Cross model:

$$\eta_R = \frac{1}{1 + (K\,\dot{\gamma}_R)^n}\,. \tag{5}$$

The temperature dependence is included in the zero-shear viscosity using the Arrhenius law:

$$\eta_0(\vartheta) = A \exp\left(\frac{B}{\vartheta}\right), \tag{6}$$

where ϑ is absolute temperature. Therefore

$$\eta(\dot{\gamma}, \vartheta) = \frac{\eta_0(\vartheta)}{1 + (K\eta_0(\vartheta)\dot{\gamma})^n}\,. \tag{7}$$

The value of the parameters are

$$\left.\begin{array}{ll} A = 1.67 \times 10^{-6}\ \mathrm{Pa\,s}\,, & B = 11174^\circ\mathrm{K}\,, \\ K = 8.55 \times 10^{-7}\ \mathrm{Pa}^{-1}\,, & n = 0.94\,. \end{array}\right\} \tag{8}$$

3. Formulation of the Problem

We consider steady, incompressible, non-isothermal flow in a capillary. The capillary geometry is shown in Fig. 3. The computational geometry, Ω, is made up of two

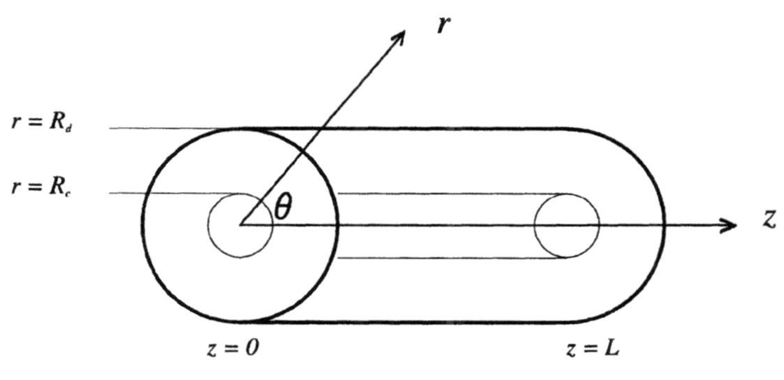

Fig. 3. Polar co-ordinate description of the capillary geometry.

domains: the flow domain, Ω_f, and the die wall, Ω_d. We denote by Γ the interface between the two domains. The natural choice of co-ordinate system is cylindrical polars (r, θ, z), with the origin taken to be at the intersection of the capillary axis and entrance. The fluid enters the geometry at $z = 0$ and exits at $z = L$. The radius of the capillary is given by $r = R_c$ and the radius of the die by $r = R_d$. The equations of motion for a generalized Newtonian fluid are:

MASS

$$\nabla . \underline{v} = 0\,, \tag{9}$$

MOMENTUM

$$\varrho(\underline{v}\,.\nabla)\underline{v} + \nabla p - \nabla.\,\underline{\underline{T}} - \varrho\underline{g} = \underline{0}\,, \tag{10}$$

ENERGY

$$\varrho C_P(\underline{v}\,.\nabla)\vartheta - k_f\nabla^2\vartheta - \underline{\underline{T}} : \nabla\underline{v} = 0\,, \tag{11}$$

EXTRA-STRESS

$$\underline{\underline{T}} = 2\,\eta(\dot{\gamma},\vartheta)\,\underline{\underline{d}}\,, \tag{12}$$

where $\underline{v}$ is the velocity vector, p is pressure, ϱ is the fluid density, C_P is the specific heat capacity, k_f is the thermal conductivity of the fluid and the rate of deformation tensor $\underline{\underline{d}} = \frac{1}{2}\left(\nabla\underline{v} + \nabla\underline{v}^{\dagger}\right)$. The viscosity depends on both the shear-rate and temperature. The first term in the energy equation describes the transfer of heat by convection, the second describes the transfer of heat by conduction and the third is the viscous dissipation term. Since we consider a generalized Newtonian model which is by definition inelastic, none of the energy imparted to the fluid by the work done by the stresses is stored. Instead it is converted irreversibly into heat by viscous dissipation. The dependence of viscosity on temperature leads to a coupling of the momentum and energy equations.

Since there is no azimuthal contribution to the flow ($\partial/\partial\theta \equiv 0$) it can be considered to be two-dimensional, with velocity vector $\underline{v} = (u, 0, w)$ and the capillary axis, $r = 0$, is an axis of symmetry. The geometry considered is therefore that shown in Fig. 4. Since the extra-stress tensor is defined explicitly by (12), it can be eliminated from the equations by substituting it into (10) and (11). We note that since gravity acts in the axial direction $\underline{g} = (0, 0, g)$.

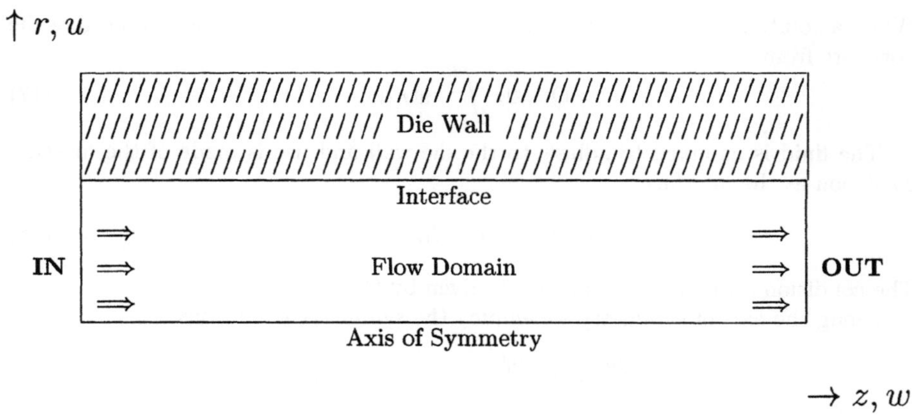

Fig. 4. Schematic representation of the two-dimensional capillary geometry.

The non-isothermal flow of the fluid is coupled with heat conduction in the solid die wall, which is described by the heat conduction equation:

$$\nabla^2\vartheta = 0\,. \tag{13}$$

At the interface of the fluid and die wall heat is transferred by conduction and we impose the continuity of heat flux:

$$(\underline{q} \cdot \underline{n})_{\text{fluid}} = (\underline{q} \cdot \underline{n})_{\text{die}} , \tag{14}$$

where $\underline{q} = -k\nabla\vartheta$ is the heat flux vector and $\underline{n} = (1, 0, 0)$ is the outward normal to the interface. Since heat is generated within the fluid, (14) states that the flux of heat exiting the fluid must be equal to the flux of heat entering the die wall.

The governing equations (9) – (13) with the flux condition (14), describe the non-isothermal flow of a generalized Newtonian fluid in a capillary coupled with heat conduction in the die wall. To complete the specification of the problem appropriate boundary conditions on the dependent variables are required.

3.1. Flow Boundary Conditions

The fluid is assumed to enter the capillary in fully-developed, isothermal flow:

$$u = 0 \, , \; \frac{\partial w}{\partial z} = 0 \, , \; \vartheta = \vartheta_0 \, , \qquad 0 < r < R_c \, , \; z = 0 \, , \tag{15}$$

where ϑ_0 is the initial temperature of the fluid (the operating temperature).

We consider the flow to be driven by a pressure gradient which is specified by the conditions:

$$\begin{aligned} p &= P_{in} \, , \qquad 0 < r < R_c \, , \; z = 0 \, , \\ \text{and} \quad p &= 0 \, , \qquad 0 < r < R_c \, , \; z = L \, . \end{aligned} \tag{16}$$

When a solution of the non-isothermal flow is obtained we can then determine the flow-rate from

$$Q = 2\pi \int_0^{R_c} w \, r \, \mathrm{d}r \, . \tag{17}$$

The fluid is assumed to adhere to the die wall and use is made of the 'no-slip' condition at the interface:

$$u = w = 0 \, , \qquad r = R_c \, , \; 0 \le z \le L \, . \tag{18}$$

The condition on ϑ at this boundary is given by (14).

Along the axis of symmetry we employ the symmetry conditions:

$$u = 0 \, , \; \frac{\partial w}{\partial r} = 0 \, , \; \frac{\partial \vartheta}{\partial r} = 0 \, , \quad r = 0 \, , \; 0 \le z \le L \, . \tag{19}$$

It is normally assumed that the fluid exits the capillary in fully developed flow, both mechanically and thermally. However, the capillaries used in practice are rarely long enough to achieve a fully developed temperature field. The temperature field remains in a state of thermal development up to the exit. Since temperature development affects the velocity field through the temperature-dependent viscosity, a fully-developed velocity field cannot be presupposed. In this study no explicit conditions on the velocity at the exit are imposed. The velocity can be inferred from the imposition of the condition $p = 0$.

The transfer of heat at the exit is extremely complex since both conduction and convection are involved. The situation can be described by defining a *heat transfer coefficient*, h, from Newton's 'law of cooling'[4]:

$$q = h(\vartheta - \vartheta_0), \tag{20}$$

where q is the heat flux at the boundary. The heat transfer coefficient of the fluid will be denoted by h_f and will be taken as constant for the flow. The boundary condition on ϑ can then be written as

$$\frac{\partial \vartheta}{\partial z} = -\frac{h_f}{k_f}(\vartheta - \vartheta_0),$$

or

$$\frac{\partial \vartheta}{\partial z} = -H_f(\vartheta - \vartheta_0), \qquad 0 < r < R_c, \; z = L. \tag{21}$$

The heat transfer coefficient is not solely dependent on the material since the geometry, flow velocity and temperature distribution can all have an effect. Therefore values of h are not commonly tabulated. Due to this lack of data we shall employ a range of values of h_f, and thus H_f, and show how the flow depends on the choice of this parameter.

3.2. Die Wall Boundary Conditions

Along the boundaries of the die wall, conditions on the temperature need only be specified. At $z = 0$ the surface of the die wall is in direct contact with the reservoir of fluid in the viscometer barrel, which is at the operating temperature, and it is therefore assumed that

$$\vartheta = \vartheta_0, \qquad R_c < r < R_d, \; z = 0. \tag{22}$$

Along $r = R_d$, the surface of the die wall is maintained at the constant operating temperature by thermostatic control, so we have

$$\vartheta = \vartheta_0, \qquad r = R_d, \; 0 \le z \le L. \tag{23}$$

The boundary condition at the die wall surface $z = L$ can be described in a similar way as the exit condition for the fluid:

$$\frac{\partial \vartheta}{\partial z} = -H_d(\vartheta - \vartheta_0), \qquad R_c < r < R_d, \; z = L, \tag{24}$$

where $H_d = h_d/k_d$, h_d is the heat transfer coefficient of the die wall material and k_d is the thermal conductivity of the die wall material. Again we shall employ a range of values of H_d and show the effect the choice of this parameter has on flow.

3.3. The Biot Condition

An alternative formulation of the problem replaces the heat conduction problem in the die wall by an appropriate boundary condition at the interface. The condition we use can be derived from a simple one-dimensional analysis of the heat conduction problem in the die wall[14].

If axial heat conduction in the die wall is neglected, (13) can be re-written, in polar co-ordinate form, as

$$\frac{\partial}{\partial r}\left(r\frac{\partial\vartheta}{\partial r}\right) = 0\,. \tag{25}$$

If the temperature at the interface is then assumed to be given by ϑ_i, (25) has the two boundary conditions

$$\vartheta = \vartheta_i \quad \text{at} \quad r = R_c$$

$$\text{and} \quad \vartheta = \vartheta_0 \quad \text{at} \quad r = R_d\,. \tag{26}$$

The radial temperature gradient in the wall can then be shown to be

$$\frac{\partial\vartheta}{\partial r} = \frac{1}{r}\frac{(\vartheta_i - \vartheta_0)}{\ln(R_c/R_d)}\,, \qquad R_c \le r \le R_d\,, \tag{27}$$

so that the flux of heat entering the die wall at the interface is given by

$$\left(-k_d\frac{\partial\vartheta}{\partial r}\right)_{\text{die}} = -\frac{k_d}{R_c}\frac{(\vartheta_i - \vartheta_0)}{\ln(R_c/R_d)}\,. \tag{28}$$

From the condition of continuity of heat flux at the interface, (14), we then have

$$\left(-k_f\frac{\partial\vartheta}{\partial r}\right)_{\text{fluid}} = -\frac{k_d}{R_c}\frac{(\vartheta_i - \vartheta_0)}{\ln(R_c/R_d)}\,, \tag{29}$$

so the temperature gradient in the fluid at the die wall is given by

$$\frac{\partial\vartheta}{\partial r} = -\frac{1}{R_c}\frac{k_d}{k_f}\frac{(\vartheta_i - \vartheta_0)}{\ln(R_d/R_c)}\,. \tag{30}$$

The condition (30) has been identified by Winter[22] as the *Biot condition*:

$$\frac{\partial\vartheta}{\partial r} = -\frac{\text{Bi}}{R_c}(\vartheta_i - \vartheta_0)\,, \tag{31}$$

where the Biot number

$$\text{Bi} = \frac{k_d}{k_f}\frac{1}{\ln(R_d/R_c)}\,, \tag{32}$$

is a measure of the ability of the die wall material to conduct heat from the fluid to the surroundings.

In practice, the wall temperature ϑ_i is unknown and is generally dependent on axial position. Therefore the Neumann condition (31) becomes the Robbins condition:

$$\frac{\partial\vartheta}{\partial r} = -\frac{\text{Bi}}{R_c}(\vartheta - \vartheta_0)\,, \qquad r = R_c\,,\ 0 \le z \le L\,. \tag{33}$$

This is the boundary condition on the capillary wall for the *Biot Problem*.

The number Bi can be determined *a priori* since it is expressed in terms of the geometric dimensions and thermal conductivities: no knowledge of the temperature field is required. We also note the two limiting cases of the Biot condition. If $\text{Bi} = 0$, we have the insulated wall condition, $\partial\vartheta/\partial r = 0$, and as $\text{Bi} \to \infty$ we obtain the isothermal wall condition, $\vartheta = \vartheta_0$.

3.4. Dimensionless Numbers

The governing equations will be solved in dimensional form. However, it is useful to define dimensionless numbers that characterize the flow[15]. Non-dimensionalization requires length, velocity and temperature scales, which provide standards to which all other measurements can be compared. We employ the length scales R_c, in the r-direction, and L, in the z-direction, and velocity scale

$$W = \frac{Q}{\pi R_c^2} \,. \tag{34}$$

The temperature scale is usually taken to be a characteristic temperature difference of the flow: $(\Delta\vartheta)_c$. However, non-isothermal flows that are due to viscous dissipation do not possess a characteristic temperature difference. We therefore use the temperature sensitivity of the viscosity:

$$\beta = \left(-\frac{1}{\eta}\frac{\partial \eta}{\partial \vartheta}\right)_{\bar{\dot{\gamma}},\vartheta_0} , \tag{35}$$

where $\bar{\dot{\gamma}} = W/R_c$ is the reference shear-rate. This gives

$$(\Delta\vartheta)_c = \frac{1}{\beta} \,. \tag{36}$$

If $\eta(\dot{\gamma}, \vartheta)$ is a separable function of shear-rate and temperature, the sensitivity of the temperature function only is required.

The relevant dimensionless numbers are:

– Nahme number:

$$\mathrm{Na} = \frac{\beta \bar{\eta} W^2}{k_f} , \tag{37}$$

where $\bar{\eta}$ is the reference viscosity:

$$\bar{\eta} = \eta(\bar{\dot{\gamma}}, \vartheta_0) \,. \tag{38}$$

The Nahme number is a measure of the extent of the coupling between the momentum and energy equations. It compares dissipation with conduction and determines whether heat generated will lead to temperature rises in the fluid sufficient to affect the velocity field. For small vales of Na, < 0.1 say, the flow can be considered to be isothermal.

– Péclet number:

$$\mathrm{Pe} = \frac{\varrho C_P W R_c}{k_f} , \tag{39}$$

which compares heat transfer by convection and heat transfer by conduction. Péclet numbers are almost always large in polymer processing since polymers are poor conductors of heat.

4. Time-Splitting Scheme

A time-dependent algorithm is used to determine the solution of the steady-state problem. Time derivatives are introduced into the momentum and energy equations. In their subsequent temporal discretization the time step plays the role of an iteration parameter, the choice of which is made to ensure rapid convergence to the steady state solution.

MOMENTUM

$$\frac{\partial \underline{\mathrm{v}}}{\partial t} + \varrho(\underline{\mathrm{v}}\,.\nabla)\underline{\mathrm{v}} + \nabla p - \nabla .\, \underline{\underline{\mathrm{T}}} - \varrho \underline{\mathrm{g}} = \underline{0}\,, \tag{40}$$

ENERGY

$$\frac{\partial \vartheta}{\partial t} + \varrho C_P(\underline{\mathrm{v}}\,.\nabla)\vartheta - k_f \nabla^2 \vartheta - \underline{\underline{\mathrm{T}}} : \nabla \underline{\mathrm{v}} = 0\,, \tag{41}$$

In this section a time-splitting scheme for the temporal discretization of the governing equations is described. The advantage of a scheme of this type is that it decouples the velocity, pressure and temperature computations. The basic scheme is accelerated to convergence using the method of false transients[16]. The scheme may thus be viewed as an iterative scheme for the corresponding steady problem. We start with initial approximations, $\underline{\mathrm{v}}^0$, p^0 and ϑ^0, to $\underline{\mathrm{v}}$, p and ϑ, respectively. After n time steps the current approximations to $\underline{\mathrm{v}}$, p and ϑ are $\underline{\mathrm{v}}^n$, p^n and ϑ^n, respectively. With knowledge of these approximations one iteration of the solution algorithm can be described. Each iteration involves three stages.

STAGE 1A

$$\frac{\underline{\mathrm{v}}^* - \underline{\mathrm{v}}^n}{\Delta_m t} = \nabla .\, \underline{\underline{\mathrm{T}}}^n - \varrho(\underline{\mathrm{v}}^n.\nabla)\underline{\mathrm{v}}^n + \varrho \underline{\mathrm{g}}, \tag{42}$$

STAGE 1B

$$\frac{\underline{\mathrm{v}}^{n+1} - \underline{\mathrm{v}}^*}{\Delta_m t} = -\nabla p^{n+1}\,, \tag{43}$$

$$\nabla .\, \underline{\mathrm{v}}^{n+1} = 0\,. \tag{44}$$

In Stage 1a an intermediate velocity field $\underline{\mathrm{v}}^*$ is determined explicitly and in Stage 1b the pressure and velocity fields are determined in such a way that $\underline{\mathrm{v}}^{n+1}$ is divergence-free.

If we take the divergence of (43) and employ (44), we obtain the Poisson equation:

$$\nabla^2 p^{n+1} = \frac{1}{\Delta_m t} \nabla .\, \underline{\mathrm{v}}^*. \tag{45}$$

Solution of the implicit equation (45) gives the new pressure field, which can be inserted into (43) to update the velocity vector $\underline{\mathrm{v}}^{n+1}$. We impose $\underline{\mathrm{v}}^{n+1} = 0$ on Γ.

STAGE 2

$$\frac{\vartheta^{n+1} - \vartheta^n}{\Delta_e t} = -\varrho C_P \left(\underline{\mathrm{v}}^{n+1}.\nabla\right) \vartheta^n + k_f \nabla^2 \vartheta^n + \underline{\underline{\mathrm{T}}}^n : \nabla \underline{\mathrm{v}}^{n+1}\,. \tag{46}$$

An explicit method is used to discretize the energy equation. This method is first-order in time. This equation is used to update the temperature in the flow field. Note that the time step in the energy equation, $\Delta_e t$, may differ from the time step in the momentum equation, $\Delta_m t$.

In the case of the *Full Problem* we aditionally solve the heat conduction equation:

$$\nabla^2 \vartheta^{n+1} = 0\,, \tag{47}$$

with the interface condition

$$\left(-k_f \frac{\partial \vartheta}{\partial r}\right)^{n+1}_{\text{fluid}} = \left(-k_d \frac{\partial \vartheta}{\partial r}\right)^{n+1}_{\text{die}}\,. \tag{48}$$

For the *Biot Problem* the boundary condition is

$$\left(\frac{\partial \vartheta}{\partial r}\right)^{n+1} = -\frac{\mathrm{Bi}}{R_c}(\vartheta^{n+1} - \vartheta_0)\,. \tag{49}$$

STAGE 3

In the final stage the shear-rate $\dot\gamma^{n+1}$ is determined from $\underline{\mathrm{v}}^{n+1}$ according to

$$\dot\gamma^{n+1} = (2\,\underline{\underline{\mathrm{d}}}^{n+1} : \underline{\underline{\mathrm{d}}}^{n+1})^{\frac{1}{2}}\,. \tag{50}$$

The viscosity over domain Ω_f and Γ can then be updated according to the chosen model:

$$\eta^{n+1} = \eta(\dot\gamma^{n+1}, \vartheta^{n+1}). \tag{51}$$

After each iteration we test for convergence of the algorithm using a normalized or relative difference:

$$e_F^{n+1} = \frac{\max|F^{n+1} - F^n|}{\max|F^{n+1}|}\,, \tag{52}$$

where F represents $\underline{\mathrm{v}}$, p or ϑ, with the requirement that

$$\max\left(e_{\underline{\mathrm{v}}}^{n+1}, e_p^{n+1}, e_\vartheta^{n+1}\right) < \varepsilon\,, \tag{53}$$

for a given tolerance ε.

5. Finite Difference Discretization

The whole geometry, in $z - r$ space, is covered with a uniform rectangular grid with M grid spacings Δz and $N_1 + N_2$ grid spacings Δr in the z and r-directions, respectively, so that we have $M + 1$ gridpoints in the z-direction and $N_1 + N_2 + 1$ gridpoints in the r-direction.

The grid consists of a set of lines parallel to the r-axis, given by

$$z_i = i\Delta z, \qquad i = 0, 1, \ldots, M\,, \tag{54}$$

and a set of lines parallel to the z-axis, given by

$$r_j = j\Delta r, \qquad j = 0, 1, \ldots, N_1 + N_2\,. \tag{55}$$

In the radial direction the number of gridpoints is chosen such that the interface is a grid-line, given by $r = r_{N_1} = N_1 \Delta r$. The point with co-ordinates $(z_i, r_j : i = 0, 1, \ldots, M,\ j = 0, 1, \ldots, N_1 + N_2)$, where the lines given by (54) and (55) intersect, is defined as the (i, j)th gridpoint. The value of the function F at the (i, j)th gridpoint is then defined as $F_{i,j}$.

The governing equations of both the *Full* and *Biot Problems*, in cylindrical co-ordinates, together with the boundary conditions, are discretized using standard second-order finite differences. The discretization of the explicit stages of the time-splitting scheme are straightforward and therefore require no comment. We describe the finite difference discretization of the pressure Poisson problem (45) and its subsequent solution.

In cylindrical polar co-ordinates, (45) can be written as

$$\frac{1}{r}\frac{\partial}{\partial r}\left(r\frac{\partial p^{n+1}}{\partial r}\right) + \frac{\partial^2 p^{n+1}}{\partial z^2} = \frac{1}{\Delta_m t}\left(\frac{1}{r}\frac{\partial (ru^*)}{\partial r} + \frac{\partial w^*}{\partial z}\right), \tag{56}$$

which, upon discretization using central differences, becomes

$$\begin{aligned}\frac{1}{\Delta z^2} p^{n+1}_{i-1,j} + \frac{(j-\frac{1}{2})}{j\Delta r^2} p^{n+1}_{i,j-1} - 2\left(\frac{1}{\Delta r^2} + \frac{1}{\Delta z^2}\right) p^{n+1}_{i,j} + \frac{(j+\frac{1}{2})}{j\Delta r^2} p^{n+1}_{i,j+1} + \frac{1}{\Delta z^2} p^{n+1}_{i+1,j} \\ = \frac{1}{\Delta_m t}\left(\frac{1}{r_j}\Delta^0_r (r_j u^*_{i,j}) + \Delta^0_z w^*_{i,j}\right),\end{aligned} \tag{57}$$

where Δ^0_r and Δ^0_z are second-order central differences, given by

$$\Delta^0_r F_{i,j} = \frac{F_{i,j+1} - F_{i,j-1}}{2\Delta r} + O(\Delta r^2) \tag{58}$$

and

$$\Delta^0_z F_{i,j} = \frac{F_{i+1,j} - F_{i-1,j}}{2\Delta z} + O(\Delta z^2), \tag{59}$$

respectively.

Equation (57) is employed at the interior gridpoints $(i, j) \in \Omega_f$. At the boundaries of domain Ω_f, and on Γ, we use the following conditions.

At the entrance we impose the pressure:

$$p^{n+1}_{i,j} = P_{in}\,. \tag{60}$$

Along the axis of symmetry we apply the symmetry condition, $\partial p / \partial r = 0$, so that,

$$\Delta^+_r p^{n+1}_{i,j} = 0\,,$$

where Δ^+_r is the standard second-order forward difference, giving

$$-3 p^{n+1}_{i,j} + 4 p^{n+1}_{i,j+1} - p^{n+1}_{i,j+2} = 0\,. \tag{61}$$

The boundary condition on p^{n+1} on Γ is obtained by projecting (43) on the outward normal to the boundary. This gives us the condition:

$$\frac{\partial p^{n+1}}{\partial r} = -\frac{(u^{n+1} - u^*)}{\Delta_m t}\,. \tag{62}$$

Since $u^{n+1} = 0$ on Γ,

$$\frac{\partial p^{n+1}}{\partial r} = \frac{u^*}{\Delta_m t}, \tag{63}$$

which can be written in discretized form as

$$\Delta_r^- p_{i,j}^{n+1} = \frac{u_{i,j}^*}{\Delta_m t}, \qquad (i,j) \in \Gamma,$$

where Δ_r^- is the standard second-order backward difference, so that

$$\frac{p_{i,j-2}^{n+1} - 4\,p_{i,j-1}^{n+1} + 3\,p_{i,j}^{n+1}}{2\,\Delta r} = \frac{u_{i,j}^*}{\Delta_m t}, \quad (i,j) \in \Gamma. \tag{64}$$

At the exit, we impose zero pressure:

$$p_{i,j}^{n+1} = 0. \tag{65}$$

The linear system of equations derived in this section can be described by the matrix equation:

$$\mathbf{A}\,\underline{\mathrm{p}}^{n+1} = \underline{\mathrm{b}}, \tag{66}$$

where $\mathbf{A}$ is a $(M+1) \times (N_1+1)$ square matrix of the coefficients of the $p_{i,j}^{n+1}$'s of equations (57), (60), (61), (64) and (65), $\underline{\mathrm{p}}^{n+1}$ is the $[(M+1) \times (N_1+1)]$th order vector of the $p_{i,j}^{n+1}$'s to be determined, and $\underline{\mathrm{b}}$ is the right hand side vector.

The equations are compiled in such a way that

$$\underline{\mathrm{p}}^{n+1} = (p_{0,0}^{n+1}, p_{0,1}^{n+1}, \ldots, p_{0,N_1}^{n+1}, p_{1,0}^{n+1}, \ldots, p_{1,N_1}^{n+1}, \ldots\ldots, p_{M,0}^{n+1}, \ldots, p_{M,N_1}^{n+1})^{\dagger}. \tag{67}$$

At the first iteration of the algorithm, the sparse matrix $\mathbf{A}$ is decomposed into upper and lower triangular forms using a sparse variant of Gaussian elimination, and the factors are stored. This decomposition is carried out using NAG routine F01BRF[25]. At each iteration we then compute $\underline{\mathrm{p}}^{n+1}$, using NAG routine F04AXF[26], by forward and back substitution. This gives us the new pressure field $\underline{\mathrm{p}}^{n+1}$ over domain Ω_f and Γ.

The governing equations and boundary conditions for ϑ are compiled in a similar manner, and the temperature is updated using a similar solution procedure as that described for the pressure.

Finally we calculate the flow-rate, Q, from the axial velocity, w, according to (17). In order to verify mass balance, we calculate the flow-rate at the entrance, Q_{en}, and at the exit, Q_{ex}:

$$Q_{\mathrm{en}} = 2\pi \int_0^{R_c} w(0,r)\, r\, \mathrm{d}r, \tag{68}$$

$$Q_{\mathrm{ex}} = 2\pi \int_0^{R_c} w(L,r)\, r\, \mathrm{d}r, \tag{69}$$

and calculate a mass balance error

$$\mathrm{MBE} = \left| \frac{Q_{\mathrm{en}} - Q_{\mathrm{ex}}}{Q_{\mathrm{en}}} \right|. \tag{70}$$

We use NAG routine E01BEF[24] to determine a piecewise cubic Hermite interpolant to the wr data and evaluate the integrals numerically using NAG routine E01BHF[24].

6. Numerical Results

6.1. Test Problem

We consider a test problem with an exact solution to demonstrate that the numerical method we have described gives rise to a convergent numerical approximation. The test problem we consider is the non-isothermal flow of a Newtonian fluid with temperature dependent viscosity proposed by Kearsley[5]. The non-uniform temperature field is due to viscous dissipation. In the analysis, the inertia term, $\varrho(\underline{v}\,.\nabla)\underline{v}$, in the momentum equation and the heat convection term, $\varrho C_P(\underline{v}\,.\nabla)\vartheta$, in the energy equation are neglected and the flow is assumed to be fully-developed. Thus, the problem is one-dimensional in the independent variable r. The axial pressure gradient P_z is assumed constant, the die wall is taken to be at a constant temperature, $\vartheta = \vartheta_0$, and the temperature dependence of viscosity is given by the approximate Arrhenius law:

$$\eta(\vartheta) = A \exp(-\alpha\vartheta)\,. \tag{71}$$

Under these assumptions the dimensionless temperature and axial velocity profiles possess the following closed form expressions:

$$\Theta(r) = \alpha\vartheta(r) = \ln\left[\frac{32\,\gamma}{(\chi r^4 + \gamma)^2}\right] \tag{72}$$

and

$$V(r) = \frac{A\,w(r)}{4R_c^2 P_z} = \frac{1}{\chi+\gamma} - \frac{r^2}{\chi r^4 + \gamma} + \frac{1}{\sqrt{\chi\gamma}}\tan^{-1}\left(\frac{\sqrt{\chi\gamma}(1-r^2)}{\gamma + \chi r^2}\right), \tag{73}$$

respectively, where the parameters χ and γ are given by

$$\chi = \frac{\alpha R_c^4 P_z^2}{4Ak_f}$$

and

$$\gamma = 16 - \chi + 4\sqrt{16 - 2\chi}\,.$$

The flow-rate is

$$Q = \frac{4\pi R_c^4 P_z}{A(\chi+\gamma)}\,. \tag{74}$$

We choose $\vartheta_0 = 0$ and $\chi = 3.5$ and obtain results on four grids, with 5×5, 9×9, 17×17 and 25×25 gridpoints, respectively. (Calculations were performed on a DECstation 5000/125 with single processor.) The isothermal solution is taken as an initial approximation to u, w and p in each case and the tolerance is $\varepsilon = 10^{-6}$. The values of $\Delta_m t$ and $\Delta_e t$ are fixed for any application of the described algorithm and are chosen to be as large as possible in order to achieve the most rapid convergence.

There is an optimum value for each time step. If either is taken to be greater than its optimum value, the solution procedure will ultimately fail since there is a restriction on the time step for the method to be stable. Also, as either time step is decreased below its optimum value, the number of iterations of the algorithm required to be performed to reach a convergent solution is increased.

There is no formal method that enables us to determine the optimum value of each time step. To determine the optimum value of $\Delta_m t$, we solve the isothermal problem for a range of values. We can then determine the optimum value of $\Delta_e t$ by solving the non-isothermal problem for the fixed value of $\Delta_m t$ and a range of values of $\Delta_e t$. On a 17×17 grid 2856 iterations are required for convergence of the algorithm in 7.0 CPU seconds. The flow becomes fully-developed towards the exit. It is in this fully-developed region that a comparison of results can be made.

We now compare the computed results against the analytic results (72) and (73) for each grid. Fig. 5 shows the comparison of Θ and V data. The dotted curve of each graph is the solution obtained for the constant viscosity fluid $\eta = A$ at the flow-rate given by (74), indicating the effect of temperature thinning. In Table I we give the optimum relative[1] time steps, number of iterations and CPU time for the computation on each grid.

TABLE I
Numerical details for the Kearsley problem.

Grid	$\Delta_m t/\Delta r\,\Delta z$	$\Delta_e t/\Delta r\,\Delta z$	Iterations	CPU time (s)
25×25	2.30×10^{-3}	0.460	5373	14.4
17×17	2.56×10^{-3}	0.256	2856	7.0
9×9	1.92×10^{-3}	0.384	907	2.1
5×5	1.60×10^{-3}	0.320	261	0.9

The accuracy of the computed solution increases with mesh refinement: for the 25×25 grid, the computed and analytic results differ by a maximum of 1.58% and 0.75% for the temperature and axial velocity profiles, respectively.

In Table II we compare the computed flow-rates, Q_{en} and Q_{ex}, with the analytic result given by (74).

Again, we see that the accuracy of the computations increases with mesh refinement: the difference between the computed and analytic flow-rate in the 25×25 grid case is 0.7%. By comparing Q_{en} and Q_{ex} we confirm mass balance for the computation on each grid.

The computed solutions can be said to compare very well to the analytic solution. The results of this test problem show that the algorithm described can be employed with confidence to our non-isothermal flow problem.

6.2. The Biot Condition

To test how well the Biot condition approximates the die wall heat conduction problem we compare the temperature fields computed using the *Full* and *Biot Problem*

[1] relative to the grid size

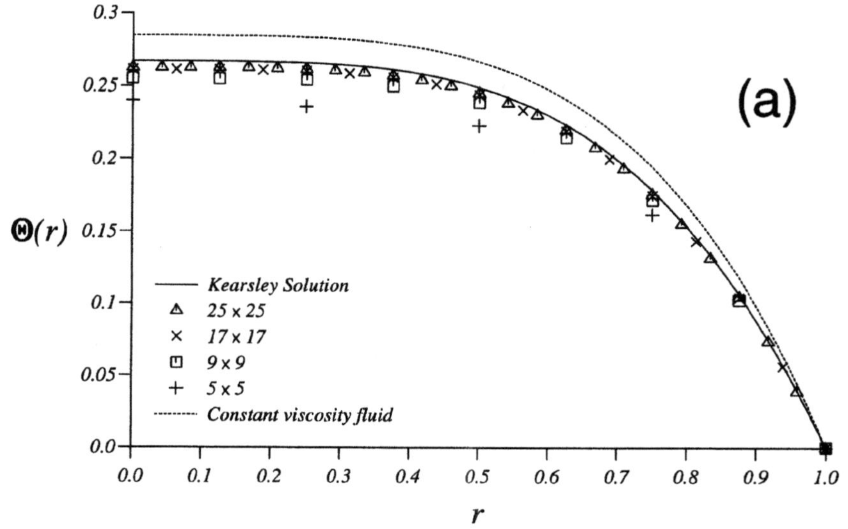

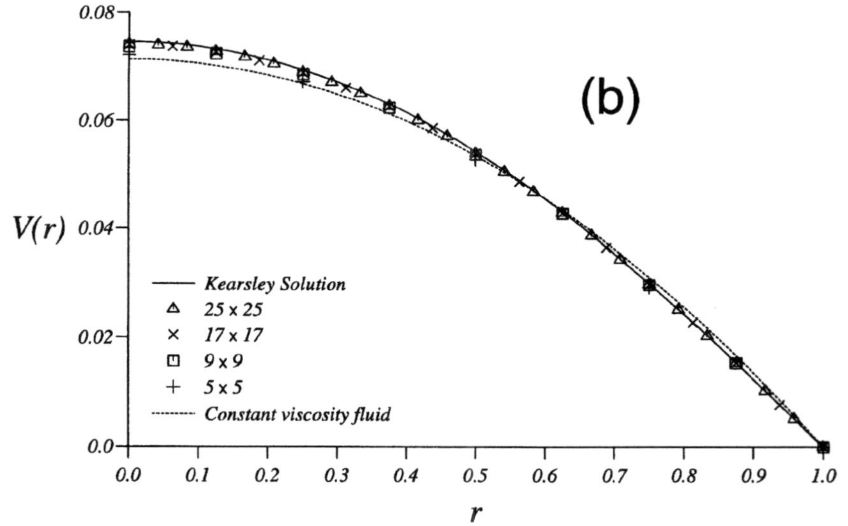

Fig. 5. (a) Temperature profiles and (b) axial velocity profiles, computed on four grids, compared with the analytic solution of Kearsley, $\chi = 3.5$.

TABLE II
Comparison of the computed flow-rates with the analytic result of the Kearsley analysis, $\chi = 3.5$, for each grid.

Analytic: $Q = 0.5309$, from (74)			
Grid	Q_{en}	Q_{ex}	MBE
25×25	0.5270	0.5272	3.79×10^{-4}
17×17	0.5265	0.5268	5.70×10^{-4}
9×9	0.5224	0.5260	6.89×10^{-3}
5×5	0.5130	0.5178	9.36×10^{-3}

formulations of the solution algorithm for the flow of the fluid described in §2.

The fluid viscosity is given by (7) with the parameters of (8). The other material parameters are $\varrho = 1200$ Kg m^{-3}, $C_P = 2300$ J Kg^{-1} °K^{-1} and $k_f = 0.215$ W m^{-1} °K^{-1}. The geometric dimensions are $R_c = 0.5$ mm and $L = 20$ mm. We take $R_d = 1$ mm, choose k_d such that Bi $= 1$ and take $\vartheta_0 = 533$°K.

We solve the Biot Problem on a grid of 25 gridpoints in the axial direction and 25 gridpoints in the radial direction, covering the flow domain and interface; and solve separately, the Full Problem on a grid of 25 gridpoints in the axial direction and $25+24$ gridpoints in the radial direction, covering the flow domain and interface (25 radial gridpoints) and the die wall (24 radial gridpoints). For both problems, the isothermal solution is used as an initial approximation and the tolerance is $\varepsilon = 10^{-6}$.

The parameter H_f is common to both problems and is chosen arbitrarily to be zero. To show the effect that the choice of H_d has on the computed temperature field, we solve the Full Problem for each of the values $H_d = 0, 10$ and 100.

Fig. 6 shows the computed temperature fields for the Biot Problem and the three Full Problems ($H_d = 0, 10$ and 100) for Pe $= 274$ and Na $= 0.52$. In Table III we give the relative time steps, iterations and CPU time for each problem.

TABLE III
Numerical details for the comparison of the Full and Biot Problems.

$\Delta_m t/\Delta r\Delta z = 2.88 \times 10^{-6}$, $\Delta_e t/\Delta r\Delta z = 2.88 \times 10^{-2}$		
Problem	Iterations	CPU time (s)
Biot (Bi $= 1$)	8055	13.7
Full ($H_d = 0$)	8054	21.0
Full ($H_d = 10$)	8054	20.7
Full ($H_d = 100$)	8054	20.5

The temperature is plotted in dimensionless form

$$\Theta = \beta\,(\vartheta - \vartheta_0)\,,$$

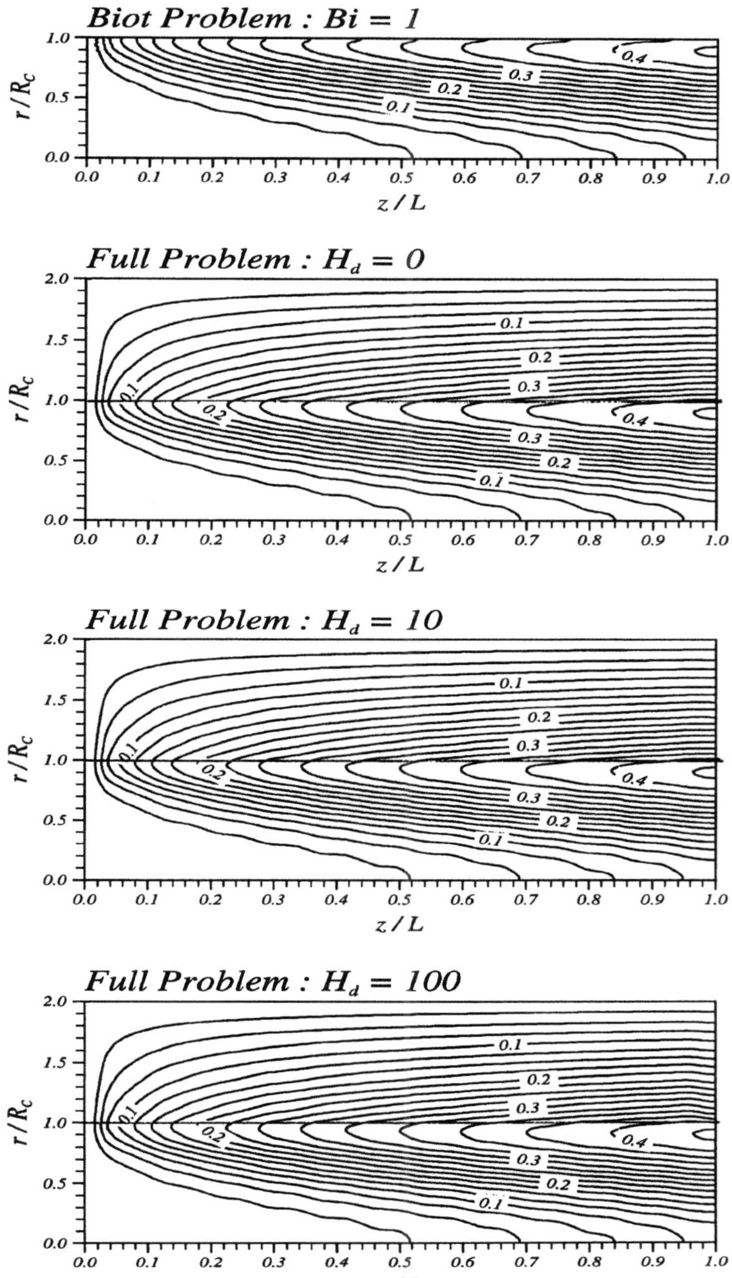

Fig. 6. Computed temperature fields, $\Theta = \beta(\vartheta - \vartheta_0)$, of the Biot Problem (Bi = 1) and Full Problem ($H_d = 0, 10, 100$) for the non-isothermal flow of Makrolon, with Pe = 274 and Na = 0.52.

where, for the viscosity model (7),

$$\beta = \frac{B}{\vartheta_0^2}\left[\frac{1+(1-n)(K\eta_0(\vartheta_0)\bar{\dot\gamma})^n}{1+(K\eta_0(\vartheta_0)\bar{\dot\gamma})^n}\right]. \tag{75}$$

For each of the four problems described above, it is found that $\beta = 0.0338\ ^\circ\mathrm{K}^{-1}$.

Radial temperature profiles at the axial positions $z/L = 1$ and $z/L = 0.75$, for each problem, are shown in Fig. 7.

In Table IV we show the computed flow-rate at the entrance and exit for each of the four problems. Also shown in Table IV is the apparent wall shear-rate:

$$\dot\gamma_a = \frac{4Q}{\pi R_c^3}, \tag{76}$$

giving an indication of the typical shear-rate of the flow. Mass balance is confirmed through comparison of Q_{en} and Q_{ex} for each problem.

TABLE IV
Computed flow-rates of the Full and Biot Problems.

	$\dot\gamma_a \approx 350\ \mathrm{s}^{-1}$		
Problem	Q_{en} $(10^{-8}\ \mathrm{m}^3\ \mathrm{s}^{-1})$	Q_{ex} $(10^{-8}\ \mathrm{m}^3\ \mathrm{s}^{-1})$	MBE
Biot (Bi = 1)	3.3490	3.3500	2.98×10^{-4}
Full ($H_d = 0$)	3.3496	3.3498	5.97×10^{-5}
Full ($H_d = 10$)	3.3496	3.3498	5.97×10^{-5}
Full ($H_d = 100$)	3.3496	3.3498	5.97×10^{-5}

The results presented for the this test case show that the Biot Problem proves to be an excellent approximation to the Full Problem in determining the temperature field in the flowing fluid (Fig. 6). By comparing the exit temperature profiles computed for the Full Problem, for H_d in the range 0 to 100 ((a) plots of Fig. 7), we see that the die wall temperature decreases with increasing H_d. However, the fluid temperature is little affected by the value of H_d over this range. Further, the temperature profiles at the axial position $z/L = 0.75$ ((b) plots of Fig. 7) show no discernible difference between the Biot Problem solution and the Full Problem solutions.

Since the heat conduction problem in the die wall does not have to be solved in the Biot Problem, the computation time required to reach a convergent solution is reduced, compared to the corresponding Full Problem. For our test case the reduction in CPU time is of the order of 30%.

As the use of the Biot condition leads to a reduction in computation time, without losing the accuracy of the Full Problem solution, the Biot Problem formulation of the solution algorithm will be employed in subsequent calculations.

7. Effects of Viscous Dissipation in Capillary Viscometry

We illustrate the effects of viscous dissipation in capillary viscometry by simulating the flow of the polycarbonate Makrolon in the experimental conditions described by

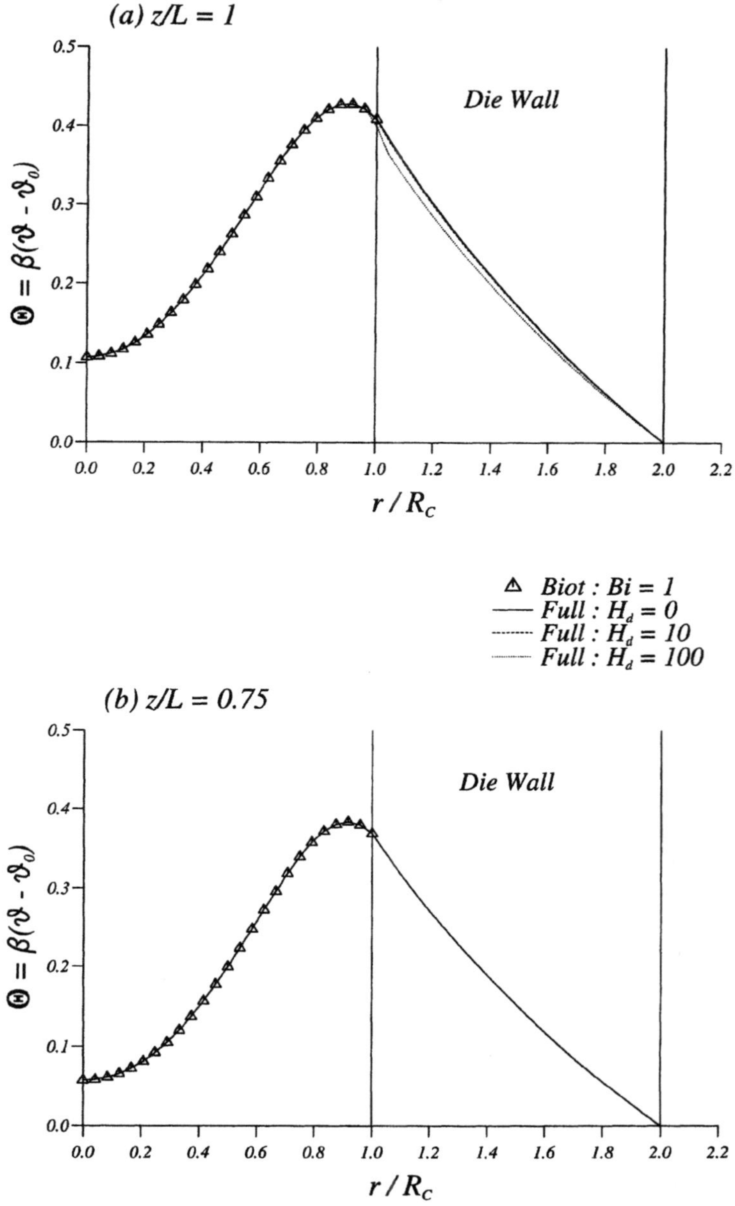

Fig. 7. Computed temperature profiles at two axial positions, (a) $z/L = 1$ and (b) $z/L = 0.75$, for the Biot Problem (Bi = 1) and Full Problem ($H_d = 0, 10, 100$) of the non-isothermal flow of Makrolon, with Pe = 274 and Na = 0.52.

Walters[27]. The viscosity determined is described in §2.

The dimensions of the capillary used in the experiment are $R_c = 0.5$ mm, $L = 20$ mm and $R_d = 9.5$ mm. The die wall material is steel which has a thermal conductivity of $k_d = 100$ W m^{-1} °K^{-1}[23], so that Bi = 158. This value of Bi tells us *a priori* that the wall temperature can be expected to show little rise in temperature.

By simulating the flow of this fluid in these experimental conditions, we shall be able to test the assumption of isothermality that was used in order to determine the viscosity originally.

We choose the operating temperature to be $\vartheta_0 = 533$°K. For this temperature, the inlet pressures of the experiment were in the range 6 – 55 MPa[28].

Fig. 8 shows the exit temperature profiles for the cases $H_f = 0, 10$ and 100, and an inlet pressure of 45 MPa. In Table V we give the computed flow-rates in

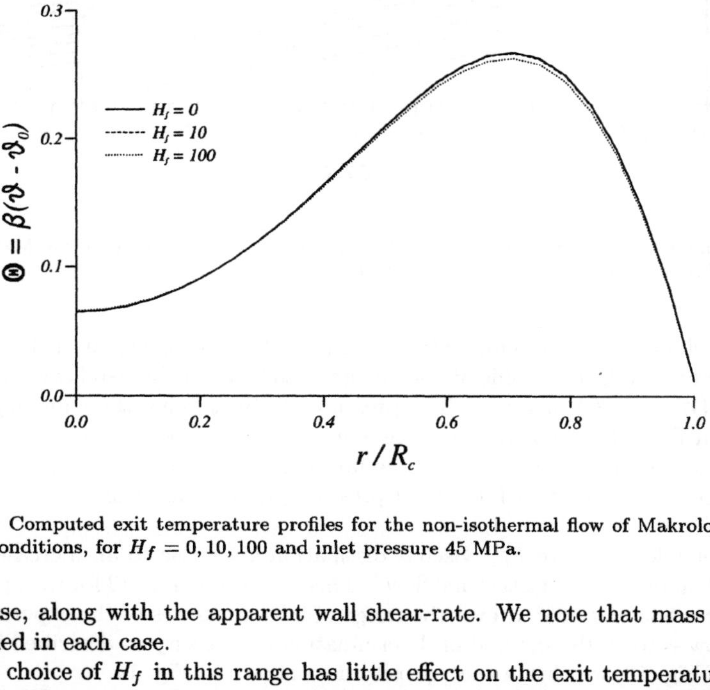

Fig. 8. Computed exit temperature profiles for the non-isothermal flow of Makrolon in experimental conditions, for $H_f = 0, 10, 100$ and inlet pressure 45 MPa.

each case, along with the apparent wall shear-rate. We note that mass balance is confirmed in each case.

The choice of H_f in this range has little effect on the exit temperature profiles and the difference in the flow-rate for each case is negligible. We shall therefore take $H_f = 0$ in subsequent computations and we shall also refer to the flow-rate Q instead of the two flow-rates Q_{en} and Q_{ex}.

The temperature field for the case $H_f = 0$ and inlet pressure 45 MPa is shown in Fig. 9 (Pe = 414, Na = 1.04 and $\beta = 0.032$ °K^{-1}), and in Fig. 10 the temperature profiles at four axial positions along the capillary are shown. We see that viscous dissipation is present; the maximum temperature increase is 8.2°K and the mean temperature increase is 4.4°K.

TABLE V
Computed flow-rates of the non-isothermal flow of Makrolon in experimental conditions, for $H_f = 0, 10, 100$, $P_{in} = 45$ MPa and $\vartheta_0 = 533°$K.

	$\dot{\gamma}_a \approx 510\ \mathrm{s}^{-1}$		
	Q_{en} (10^{-8} m^3 s^{-1})	Q_{ex} (10^{-8} m^3 s^{-1})	MBE
$H_f = 0$	5.0727	5.0731	7.88×10^{-5}
$H_f = 10$	5.0726	5.0732	5.91×10^{-5}
$H_f = 100$	5.0729	5.0730	1.97×10^{-5}

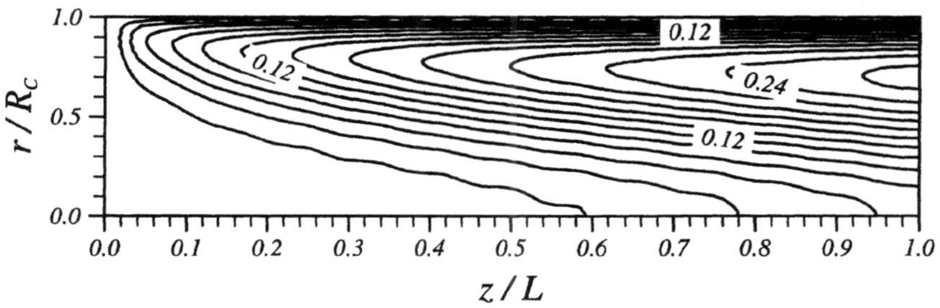

Fig. 9. Computed temperature field, $\Theta = \beta(\vartheta - \vartheta_0)$, for the non-isothermal flow of Makrolon in experimental conditions, with Pe = 414, Na = 1.04 and $\vartheta_0 = 533°$K.

The development of the temperature field leads to rearrangement of the velocity profile: the axial velocity profile at the entrance and exit of the capillary are shown in Fig. 11 ($W = 0.064$ m s^{-1}). The pressure gradient along the capillary is not constant. In Fig. 12 we plot the pressure along the axis of symmetry.

We can see from these results that the flow is non-isothermal and cannot be described as fully-developed with constant pressure gradient, as is the case in isothermal flow.

For a set inlet pressure P_{in}, viscous dissipation gives rise to an increase in the flow-rate Q, above that of isothermal flow.[2] This is shown in Fig. 13 for the operating temperatures 533 and 563°K, over the range of inlet pressures of the experiment.

If we now employ the method of determination of viscosity as described in §2 to the data of Fig. 13, we obtain the viscosity data shown in Fig. 14.

The method of reduced variables can be applied to this data, where η_0 is that described in (6). The resulting master curves are shown in Fig. 15. The upper curve corresponds to the viscosity that was the input of the direct problem and is retrieved since we solved the flow problem as isothermal. The lower curve is the viscosity determined from the results of non-isothermal flow. We see that the

[2] For readers who consider capillary flow to be driven by the imposition of a flow-rate, this is equivalent to the statement that viscous dissipation gives rise to a reduction in the recorded pressure.

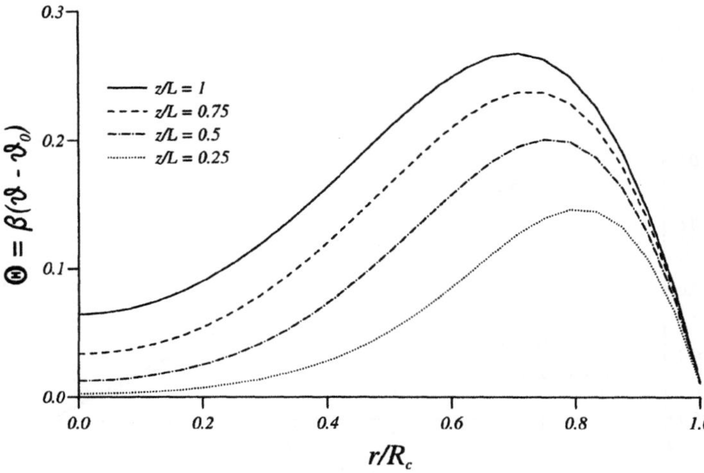

Fig. 10. Computed temperature profiles at four axial positions for the non-isothermal flow of Makrolon in experimental conditions, with Pe = 414, Na = 1.04 and $\vartheta_0 = 533°$K.

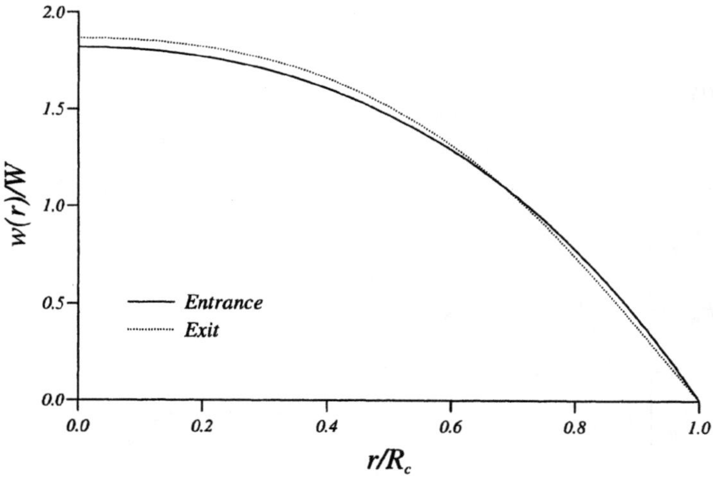

Fig. 11. Computed axial velocity profiles at the capillary entrance and exit for the non-isothermal flow of Makrolon in experimental conditions, with Pe = 414, Na = 1.04 and $\vartheta_0 = 533°$K.

viscosity determined is in error; we are unable to retrieve the input viscosity.

The maximum error at the highest shear-rates is of the order of 14%. This difference is due to the thinning effect of viscous dissipation. If viscous dissipation

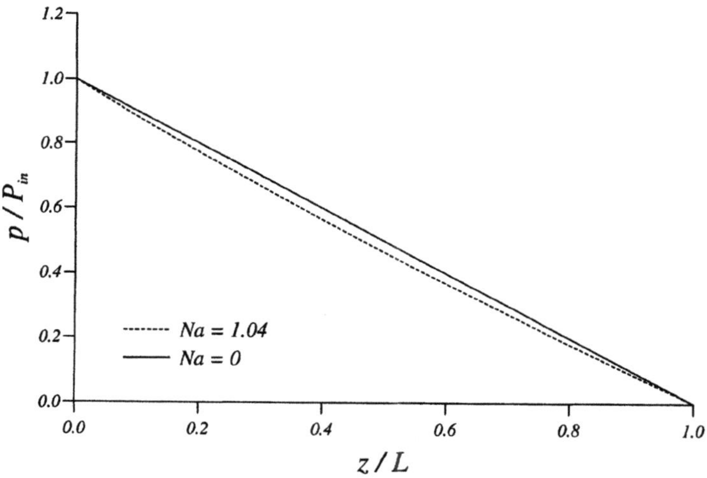

Fig. 12. Computed pressure along the axis of symmetry for the non-isothermal flow of Makrolon in experimental conditions, with Pe = 414, Na = 1.04 and $\vartheta_0 = 533°$K.

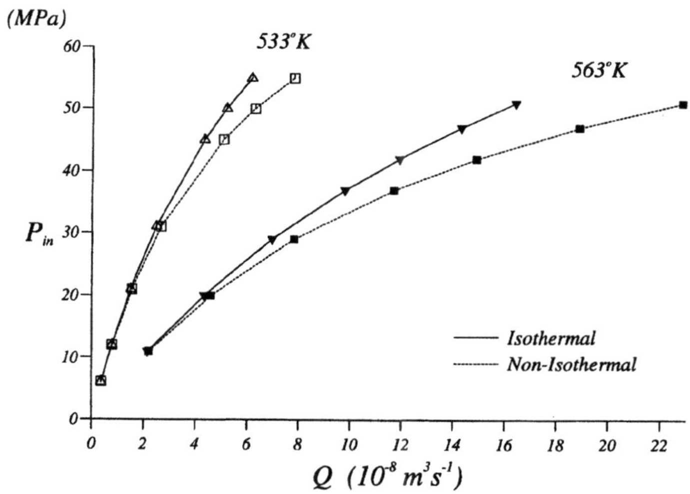

Fig. 13. Computed pressure–flow-rate curves for the non-isothermal flow of Makrolon in experimental conditions, compared to isothermal results, for the operating temperatures 533 and 563°K.

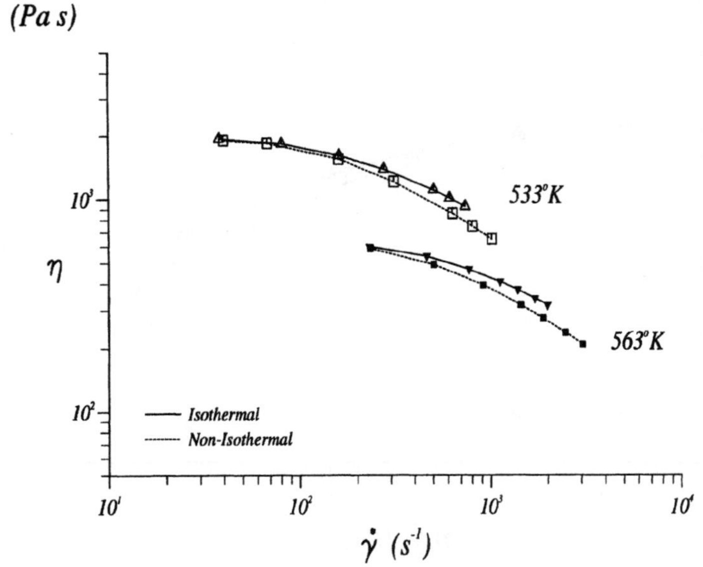

Fig. 14. Viscosity data obtained from the isothermal and non-isothermal flow data of Makrolon at the operating temperatures 533 and 563°K. The pressure–flow-rate data is shown in the previous figure.

is not taken into account when determining $\eta(\dot{\gamma})$, this thinning effect is attributed to shear thinning and we therefore obtain erroneous viscosity measurements.

From the results presented here, we can therefore assume that the flow of Makrolon, as studied by Walters[27], was non-isothermal and the viscosity determined in that study includes the effects of viscous dissipation.

8. Correction of Shear Viscosity for Viscous Dissipation Effects

In the previous section the effects of viscous dissipation in capillary viscometry were studied. The direct problem of determining the flow-rate Q for a given inlet pressure P, operating temperature ϑ_0 and viscosity model $\eta(\dot{\gamma}, \vartheta)$ was solved. It was shown that when viscous dissipation is appreciable, the conventional method of viscosity determination (see §2) fails. The viscosity determined using this analysis includes shear thinning and viscous dissipation effects. Both effects act to decrease the fluid viscosity and both give rise to an increase in the flow-rate for a given inlet pressure and operating temperature.

An alternative method has to be proposed. The problem before us is: how do we determine $\eta(\dot{\gamma}, \vartheta)$ from (P, Q, ϑ_0) data when viscous dissipation is taken into account? This is the statement of an inverse problem since the (P, Q, ϑ_0) data

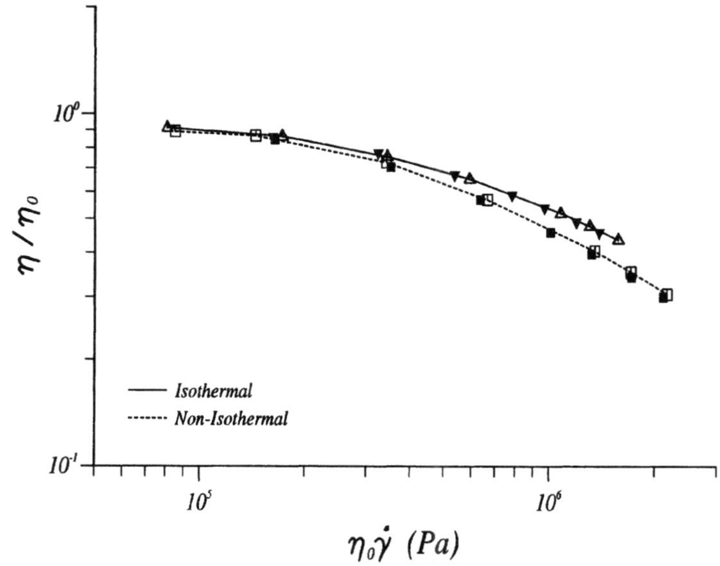

Fig. 15. Viscosity master curves obtained from the isothermal and non-isothermal flow data of Makrolon. The original viscosity–shear-rate data is shown in the previous figure. Both curves can be described by the Cross model. Solid curve: $K = 8.55 \times 10^{-7}$ Pa^{-1}, $n = 0.94$. Dotted curve: $K = 11.0976 \times 10^{-7}$ Pa^{-1}, $n = 0.9616$.

determined experimentally is clearly dependent on the viscosity of the fluid considered. The assumptions made in isothermal flow, *i.e.* the flow is fully-developed with constant axial pressure gradient, are no longer justified. The viscosity has to be determined from the solution of the complete set of governing equations of non-isothermal flow.

In this section a method of solution is proposed. We aim to correct the experimentally determined viscosity for viscous dissipation effects by correcting the viscosity model parameters that describe $\eta(\dot{\gamma}, \vartheta)$.

To illustrate our correction method we consider those fluids whose viscosity can be described by a master curve using the method of reduced variables, and without loss of generality we consider the Cross model. The viscosity can then be corrected for viscous dissipation effects by correcting the master curve parameters.

The experimentally determined viscosity is denoted by

$$\eta_{\text{exp}}(\dot{\gamma}, \vartheta) = \frac{\eta_0(\vartheta)}{1 + (K_e \eta_0(\vartheta)\dot{\gamma})^{n_e}}, \tag{77}$$

which includes the viscous dissipation effects. We strive to determine the true viscosity, without viscous dissipation effects:

$$\eta_t(\dot{\gamma}, \vartheta) = \frac{\eta_0(\vartheta)}{1 + (K_t \eta_0(\vartheta)\dot{\gamma})^{n_t}}, \tag{78}$$

by correcting the parameters K and n. We have inherently assumed that $\eta_t(\dot{\gamma}, \vartheta)$ can be described by the same viscosity model as $\eta_{\text{exp}}(\dot{\gamma}, \vartheta)$ and that the parameters that define $\eta_0(\vartheta)$ can remain uncorrected since they are determined at low shear-rates where viscous dissipation is negligible.

8.1. Least Squares Method

We have illustrated how solving a series of direct problems for a given viscosity $\eta(\dot{\gamma}, \vartheta)$ can lead to the interpretation of a different viscosity curve if viscous dissipation is not taken into account (see Fig. 15). This can be thought of as solving a series of direct problems for a given K and n, and determining a different K and n, which we shall denote by $\overline{K}$ and $\overline{n}$. For example, if we refer to Fig. 15, we see that for the input $K = 8.55 \times 10^{-7}$ and $n = 0.94$ we determine $\overline{K} = 11.0976 \times 10^{-7}$ and $\overline{n} = 0.9616$.

If we solve the direct problems for a range of (K, n), we can determine a corresponding $(\overline{K}, \overline{n})$ in each case. Thus, we can build a database of $(K, n, \overline{K}, \overline{n})$ values. We then ask the question: given $\overline{K}$ and $\overline{n}$, can we determine the corresponding K and n? That is, having solved a number of *direct problems* (finding $\overline{K}$ and $\overline{n}$ for a given K and n) can we now solve the *inverse problem* of determining K and n when we are given $\overline{K}$ and $\overline{n}$?

In an attempt to answer this question we assume that $\overline{K}$ and $\overline{n}$ can be expressed as polynomials in K and n as follows:

$$\overline{K} = \overline{K}(K, n) = \sum_{i=0}^{M} \sum_{j=0}^{M} a_{ij}\, K^i\, n^j\,, \tag{79}$$

$$\overline{n} = \overline{n}(K, n) = \sum_{i=0}^{M} \sum_{j=0}^{M} b_{ij}\, K^i\, n^j\,. \tag{80}$$

The coefficients in these expansions, a_{ij} and b_{ij}, are determined by the method of least squares. To this end, we define the error functions

$$E = \sum_{r=1}^{R} \left(\overline{K}(K_r, n_r) - \overline{K}_r\right)^2 \tag{81}$$

and

$$F = \sum_{r=1}^{R} \left(\overline{n}(K_r, n_r) - \overline{n}_r\right)^2\,, \tag{82}$$

where R is the number of $(K, n, \overline{K}, \overline{n})$ datapoints.

We minimize E by solving the set of $(M+1)^2$ linear equations:

$$\frac{\partial E}{\partial a_{pq}} = 0, \qquad p, q = 0, 1, \ldots, M\,, \tag{83}$$

for the $(M+1)^2$ unknowns a_{pq} ; $p, q = 0, 1, \ldots, M$,
i.e. we solve

$$\sum_{r=1}^{R} \left(K_r^p\, n_r^q \left[\sum_{i=0}^{M} \sum_{j=0}^{M} a_{ij}\, K_r^i\, n_r^j - \overline{K}_r \right] \right) = 0\,; \quad p, q = 0, 1, \ldots, M\,, \tag{84}$$

for a_{ij} ; $i,j = 0,1,\ldots,M$, given the data $(K,n,\overline{K},\overline{n})_r$; $r = 1,2,\ldots,R$.

We similarly minimize F by solving the linear system:

$$\sum_{r=1}^{R}\left(K_r^p\, n_r^q \left[\sum_{i=0}^{M}\sum_{j=0}^{M} b_{ij}\, K_r^i\, n_r^j - \overline{n}_r\right]\right) = 0\,; \quad p,q = 0,1,\ldots,M\,, \tag{85}$$

for b_{ij} ; $i,j = 0,1,\ldots,M$, given the data $(K,n,\overline{K},\overline{n})_r$; $r = 1,2,\ldots,R$.

Now, given the experimentally determined $\overline{K} = \overline{K}_{\mathbf{exp}}$ and $\overline{n} = \overline{n}_{\mathbf{exp}}$ and the a_{ij}'s and b_{ij}'s, we solve the simultaneous equations:

$$\left.\begin{aligned} \overline{K}\,(K,n) &= \overline{K}_{\mathbf{exp}}\,, \\ \overline{n}\,(K,n) &= \overline{n}_{\mathbf{exp}}\,, \end{aligned}\right\}$$

i.e.

$$\begin{aligned} \sum_{i=0}^{M}\sum_{j=0}^{M} a_{ij}\, K^i\, n^j &= \overline{K}_{\mathbf{exp}}\,, \\ \sum_{i=0}^{M}\sum_{j=0}^{M} b_{ij}\, K^i\, n^j &= \overline{n}_{\mathbf{exp}}\,, \end{aligned} \tag{86}$$

for K and n.

EXAMPLE

We choose to correct the viscosity parameters of Makrolon. We choose the operating temperatures to be 533 and 563°K, and solve the direct problem for 7 inlet pressures at each temperature. Thus, the determination each $(K,n,\overline{K},\overline{n})$ datapoint requires the solution of 14 direct problems. We have determined 12 datapoints (168 direct problems). From these data, we determine the K and n that correspond to the experimentally determined values of $\overline{K}_{\mathbf{exp}} = 8.55 \times 10^{-7}\ \mathrm{Pa}^{-1}$ and $\overline{n}_{\mathbf{exp}} = 0.94$.

To solve (84), (85) and (86), we have used the computer algebra package MAPLE. In general, there exists more than one pair (K,n) that will satisfy (86). In this case, the most feasible solution is selected. In practice, this has not presented a problem. For example, for the case $M = 1$, (86) was found to possess the two solutions:

$$(K,n) = (5.19 \times 10^{-7}, 0.8450) \tag{87}$$

and

$$(K,n) = (2.77 \times 10^{-6}, -0.1805)\,. \tag{88}$$

Clearly (88) is unrealistic and can be disregarded.

For the case $M = 1$ we have determined

$$K = 5.19 \times 10^{-7}\ \mathrm{Pa}^{-1} \quad \text{and} \quad n = 0.8450\,. \tag{89}$$

For the case $M = 2$ we have determined

$$K = 5.22 \times 10^{-7}\ \mathrm{Pa}^{-1} \quad \text{and} \quad n = 0.8471\,. \tag{90}$$

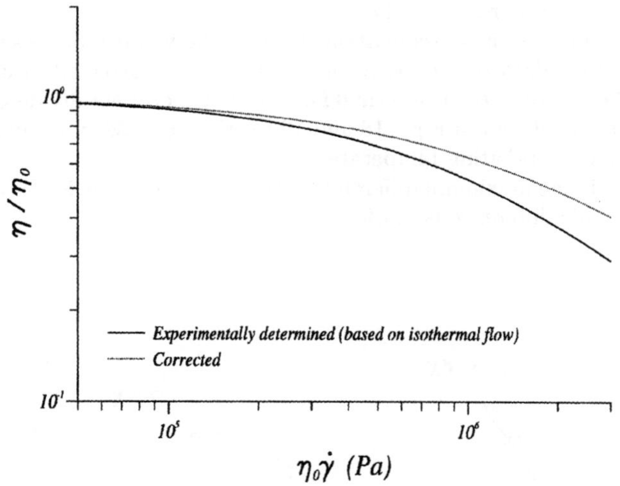

Fig. 16. Corrected viscosity curve from the least squares method ($M = 1$) compared to the experimentally determined viscosity. Solid curve: $K = 8.55 \times 10^{-7}\ \mathrm{Pa}^{-1}$, $n = 0.94$. Dotted curve: $K = 5.19 \times 10^{-7}\ \mathrm{Pa}^{-1}$, $n = 0.8450$.

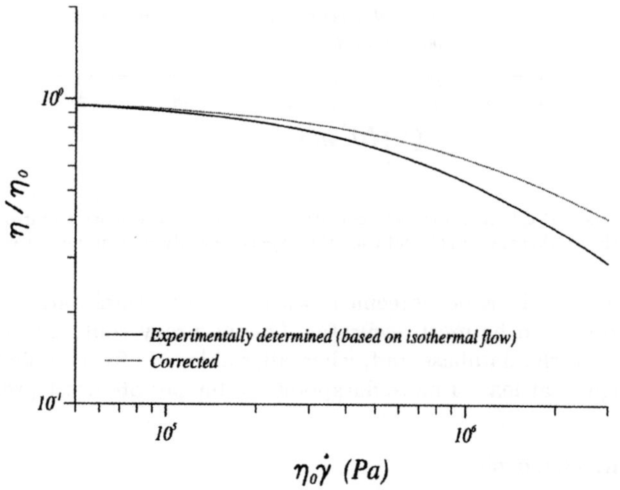

Fig. 17. Corrected viscosity curve from the least squares method ($M = 2$) compared to the experimentally determined viscosity. Solid curve: $K = 8.55 \times 10^{-7}\ \mathrm{Pa}^{-1}$, $n = 0.94$. Dotted curve: $K = 5.22 \times 10^{-7}\ \mathrm{Pa}^{-1}$, $n = 0.8471$.

The corrected viscosity curves are shown in Figs. 16 and 17, for the cases $M = 1$ and $M = 2$, respectively. At the highest shear-rates the curves differ by 25.1% and 24.8% for $M = 1$ and $M = 2$, respectively.

To prove that the method is self-consistent, we test the validity of these results by comparing the (P, Q) data of the direct problems for the corrected viscosity in each case, to the (P, Q) data of the isothermal flow of the experimentally determined viscosity. The results are shown in Figs. 18 and 19 for the cases $M = 1$ and $M = 2$, respectively, and the two operating temperatures.

For the case $M = 1$, the maximum difference between the curves is 5.2% (563°K). For the case $M = 2$, this difference is 4.5%.

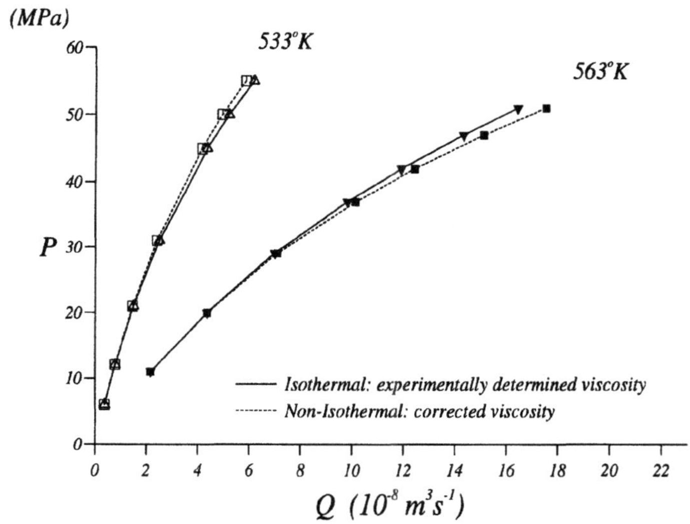

Fig. 18. Non-isothermal (P, Q) data for the corrected viscosity of the least squares method ($M = 1$), compared to the isothermal (P, Q) data for the experimentally determined viscosity.

Both can be said to be in good agreement with the isothermal data, with the result for $M = 2$ proving to be most accurate. This accuracy could be improved by increasing the size of the database and, when allowable, M. To be able to take $M = 3$, we would require at least 4 more datapoints: a further 56 direct problems.

9. Conclusions and Discussion

In this paper we have described a numerical algorithm for the solution of the governing equations of non-isothermal flow. This method has been applied to the problem of non-isothermal flow in a capillary. The governing equations and boundary conditions have been discretized using finite differences.

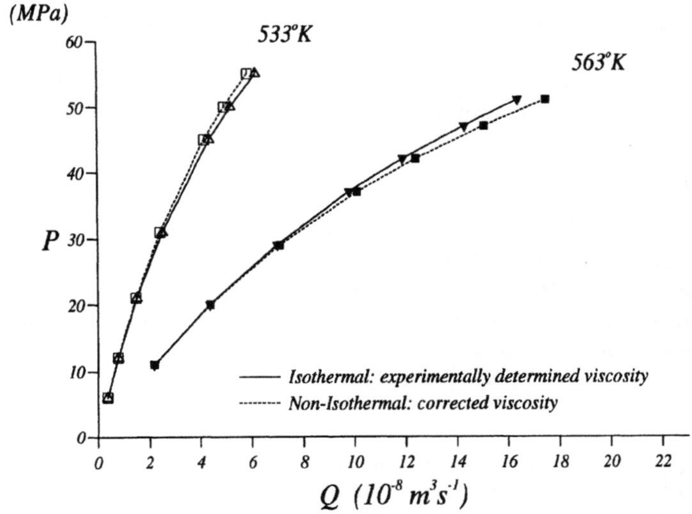

Fig. 19. Non-isothermal (P, Q) data for the corrected viscosity of the least squares method $(M = 2)$, compared to the isothermal (P, Q) data for the experimentally determined viscosity.

The validity of the algorithm was established by applying it to a test problem that possesses an analytic solution. It was then used to determine how well the Biot condition approximated the die wall heat conduction problem by comparing the solutions of the Biot and Full Problems. The solutions were seen to be in excellent agreement, giving credibility to the Biot condition as the choice of approximating thermal boundary condition for non-isothermal capillary flow that is due to viscous dissipation.

The reduction in CPU time in using the Biot Problem formulation against that of the Full Problem formulation, meant that the Biot Problem formulation was used as the solution method for the analysis of viscous dissipation in capillary viscometry.

By simulating the flow of a polymer melt in experimental conditions, the effects of viscous dissipation were shown. The conventional method of viscosity determination was seen to be unsuccessful in determining viscosity in flows with viscous dissipation: the effect of temperature thinning is attributed to shear thinning, giving erroneous viscosity measurements.

In the light of this, alternative methods of viscosity determination are required. We have described a method that may be employed to determine shear viscosity taking into account viscous dissipation. The method determines the true viscosity (without viscous dissipation effects) by correcting the parameters of the experimentally determined viscosity model.

This least squares method offers an efficient means of determining the true viscosity using inverse-problem methodology.

We limited the discussion to viscosity master curves of Cross model type. This was done to make the description of the method clear. The least squares method could be equally well applied to any general viscosity model with any number of parameters.

Although we have considered only one test case, the software is now in place and will clearly be of interest and potential use to capillary viscometer manufacturers.

Acknowledgements

The first author would like to acknowledge the United Kingdom Science and Engineering Research Council (SERC) for the award of a CASE studentship with Unilever plc. We wish to thank Professor Ken Walters for suggesting this problem and for many helpful remarks. The experimental results for Makrolon are reproduced with the permission of Mary Walters.

References

1. Brinkman, H.C., *Appl. Sci. Res.*, **A2**, 120 (1951)
2. Bird, R.B., *J. Soc. Plastics Eng.*, **11**, 35 (1955)
3. Gee, R.E. and Lyon, J.B., *I.E.C. Fund.*, **49**, 956 (1957)
4. Bird, R.B., Stewart, W.E. and Lightfoot, E.N., *"Transport Phenomena"*, Wiley & Sons (1960)
5. Kearsley, E.A., *Trans. Soc. Rheol.*, **6**, 253 (1962)
6. Gerrard, J.E. and Philippoff, W., in *"Proceedings of the Fourth International Congress on Rheology"*, 77 (1963)
7. Marshall, D.E., Klein, I. and Uhl, R.H., *S.P.E.J.*, **20**, 329 (1964)
8. Vinogradov, G.V. and Malkin, A.Ya., *J. Poly. Sci.*, **A2**, 2357 (1964)
9. Gerrard, J.E., Steidler, R.E. and Appeldoorn, J.K., *I.E.C. Fund.*, **4**, 332 (1965)
10. Gerrard, J.E., Steidler, R.E. and Appeldoorn, J.K., *I.E.C. Fund.*, **5**, 260 (1966)
11. van Leeuwen, J., *Poly. Eng. Sci.*, **7**, 98 (1967)
12. Martin, B., *Int. J. of Non-Linear Mech.*, **2**, 285 (1967)
13. Kim, H.T. and Collins, E.A., *Poly. Eng. Sci.*, **11**, 83 (1971)
14. Eckert, E.R.G. and Drake, R.M., *"Analysis of Heat and Mass Transfer"*, McGraw-Hill (1972)
15. Pearson, J.R.A., in *"Progress in Heat and Mass Transfer"*, Volume 5, Pergamon Press (1972)
16. Mallinson, G.D. and de Vahl Davis, G., *J. Comp. Phys.*, **12**, 435 (1973)
17. Cox, H.W. and Makosko, C.W., *A.I.Ch.E.J.*, **20**, 785 (1974)
18. Walters, K., *"Rheometry"*, Chapman and Hall (1975)
19. Galili, N., Takserman-Krozer, R. and Rigbi, Z., *Rheol. Acta.*, **14**, 550 (1975)
20. Galili, N., Takserman-Krozer, R. and Rigbi, Z., *Rheol. Acta.*, **14**, 816 (1975)
21. Heiber, C.A., *Rheol. Acta.*, **16**, 553 (1977)
22. Winter, H.H., *Adv. Heat Transfer*, **13**, 205 (1977)
23. Weast, R.C., *"CRC Handbook of Chemistry and Physics"*, 68th Edition, CRC Press (1988)
24. *NAG Fortran Library*, Mark 15, Volume 3
25. *NAG Fortran Library*, Mark 15, Volume 5
26. *NAG Fortran Library*, Mark 15, Volume 6
27. Walters, M.R., *"Rheometry of Polymer Melts"*, MPhil. thesis, University of Wales (1990)
28. Walters, M.R., private communication (1991)

NUMERICAL SIMULATION OF FILLING PROBLEMS RELATED TO INJECTION MOULDING

D. DING, P. TOWNSEND and M.F. WEBSTER
University of Wales Institute of non-Newtonian Fluid Mechanics,
Department of Computer Science,
University College Swansea, SA2 8PP, United Kingdom.

Abstract. Various aspects of numerical simulation for filling problems related to injection moulding are addressed. Two front tracking techniques are studied, one based on particle tracking and the other on source propagation. Both isothermal and non-isothermal processing conditions are adopted whilst modelling the injection phase only, of the complete industrial process. Some model axisymmetric shaped moulds are considered in this early stage of a wider study aimed at fully three-dimensional predictions for multi-layered thin-walled products in complex moulds.

Key words: injection moulding, polymer melts, front tracking, particle tracking, source propagation, non-isothermal, non-Newtonian.

1. Introduction

This paper reports on a project concerned with the numerical simulation of filling problems related to injection moulding that occur in the polymer processing industry. The motivation of this work is to develop an advanced technology to allow for multi-layered thin-walled products to be formed in one industrial processing step by injecting different polymeric materials into an axisymmetric or three-dimensional complex shaped mould simultaneously or sequentially. This necessitates the dual approach of an experimental programme combined with a computer simulation approach. Part of the work is being carried out by our partner at the Technical University of Eindhoven to extend existing equipment designs to deal with three component multi-layer injection moulding. The computer simulation part of the work is being performed at the University of Wales, Swansea to incorporate these industrial processes into a three-dimensional transient computational fluid dynamics(CFD) computer program. This involves front tracking to update the position of the flow front during the mould filling, and particle tracking to predict the deformation history of bands of material behind the flow front.

J.F. Dijksman and G.D.C. Kuiken (eds.),
IUTAM Symposium on Numerical Simulation of Non-isothermal Flow of Viscoelastic Liquids, 133-158.

In the literature, filling problems related to injection moulding are much cited under isothermal processing conditions. For example in reference [1], Rose first observed experimentally the relevant phenomena of fountain flow, which he defined as: 'the streamlines of a fluid advancing in a tube (or between parallel plates) appear like a fountain when viewed from a frame of reference moving with the flow front'. A number of approximate models have subsequently been proposed to describe fountain flow. Tadmor[2] developed a method called 'flow analysis network' based on a semi-quantitative approach to predict the position of the flow front and pressure field, where the front region was modelled as a planar stagnation flow. Recently, attention has been focused on computer simulation. Coyle et al.[3] used a finite element method to compute the history of the deformation of tracer lines as a function of time for a Newtonian flow near an advancing front between parallel plates, where a 'mushrooming' pattern was observed. This work was compared against experimental data for both Newtonian and shear-thinning fluids. Mavridis et al.[4] studied the transient behaviour of some filling problems by using a conventional Galerkin spatial discretization method and an Adams-Basforth predictor-corrector time-stepping scheme, obtaining the transient development of a Newtonian free surface flow and achieving good agreement with experiments. Vos et al.[5] studied fountain flow experimentally for a piston-driven problem, comparing results against a theoretical study by employing a steady penalty function finite element method. This piston-driven flow problem serves as both a precursor to the more complex filling problems, and as a means by which the initial flow during the preheating stage in the retaining accumulator vessel may be modelled. See on for comments on relevance to our own studies.

The work presented here shows some solutions obtained by employing a Taylor-Galerkin approach for filling problems. The scheme is a characteristic based method, models transient behaviour and offers computational efficiency [6,7]. The main emphasis in this study is to combine front tracking and particle tracking with the Taylor-Galerkin algorithm to model filling problems related to injection moulding. The complexity of such work comes not from the governing field differential equations, which are solved using a well established transient finite element integration scheme, but more from the need to calculate and continually update the position of the flow front as the mould is filled.

Two front tracking schemes are discussed in this paper, both sharing dependence on a background Eulerian mesh and local remeshing characteristics. One such scheme is a compression-expansion method based on particle tracking, whilst a second scheme is based on ideas of source propagation. Particle tracking is accomplished by employing a Runge-Kutta method to solve the ordinary differential equations for particle position considered in a strong sense. Numerical simulations for piston-driven flow problems and the filling of axisymmetric shaped moulds are presented. In conclusion we comment on the outstanding sections of this project: to investigate flows into more complicated three-dimensional moulds, more realistic non-isothermal processing conditions with variable Robbins type boundary conditions for the flow front, and a viscoelastic residual stress analysis following a decoupled solution approach. We have already published solutions for three-dimensional problems not involving moving fronts using the Taylor-Galerkin algorithm, for Newtonian isothermal flows[8] and non-Newtonian non-isothermal flows[9].

2. System of equations

The governing system of equations for viscous incompressible flow includes a description of momentum transport and conservation of mass. The primary variables involved are those of velocity $\mathbf{u}(\mathbf{x},t)$ and pressure $p(\mathbf{x},t)$ on the spatial domain at time t. With material parameters of density ρ and viscosity μ the generalised momentum and mass equations may be stated as

$$\rho \mathbf{u}_t = \nabla.(\mu \nabla \mathbf{u}) - \rho \mathbf{u}.\nabla \mathbf{u} - \nabla p$$

$$\nabla.\mathbf{u} = 0$$

In addition in the non-isothermal context there is a need to describe the transport of energy which may be given in terms of the primary variable temperature $T(\mathbf{x},t)$ in the form

$$\rho c_p T_t = \kappa \nabla^2 T - \rho c_p \mathbf{u}.\nabla T - \mu \Phi$$

where additional material parameters are the heat capacity at constant pressure c_p and the thermal conductivity κ. The final term in the energy equation is a viscous heat dissipation term, $\Phi = \frac{1}{2} I_2$, where I_2 is the second invariant of rate-of-strain tensor. Particular models for viscosity μ adopted are:

(a) a Carreau-Yasuda model which is used for fitting some polystyrene data:

$$\mu = B_2 e^{\frac{A_2}{T}} [1 + (B_1 e^{\frac{A_1}{T}} \gamma)^2]^{\frac{m-1}{2}},$$

(b) a shifted power-law used for some ABS data, and in particular to allow for yield stress behaviour:

$$\mu = \begin{cases} K(\varepsilon)^{m-1}, & \gamma \le c + \varepsilon \\ K(\gamma - c)^{m-1}, & \gamma > c + \varepsilon \end{cases}$$

$$K = B_1 e^{\frac{A_1}{T}},$$

where γ is the shear rate, m is a power-law index, A_1, A_2, B_1 and B_2 are material constants, c is a shear rate which corresponds to a yield stress and ε is a suitably small shear rate tolerance factor used to elevate the Newtonian viscosity plateau, say $O(10^{-4})$. The parameter c controls the deformation pattern of the flow, with a large value of $c=O(10^{-1})$ providing a flat velocity profile.

The boundary conditions to be adopted for current problems may be stated with reference to the primary variables on a typical axisymmetric flow domain and flow inlet sections, no-slip walls, flow symmetry lines and the flow front. In particular,

inlet Γ_1: $\mathbf{u}=(v_r,v_z)$ prescribed as $v_r=0$, $v_z=v_z(r)$,

walls Γ_2: no-slip, $v_r=v_z=0$,

symmetry Γ_4: $v_r=0$, $\tau_{rz}=0$,

front Γ_3: $(\tau - \delta p).\mathbf{s}=0$

where a boundary Γ_i is shown in Figure 1 with a superscript indicating time at n or n+1, v_r and v_z are the radial and axial velocity components respectively, τ is a stress tensor, δ is a unit tensor, $\mathbf{s}$ is outward pointing normal to the boundary Γ_3. Also on Γ_3 we can take $p = \frac{\partial p}{\partial s} = 0$.

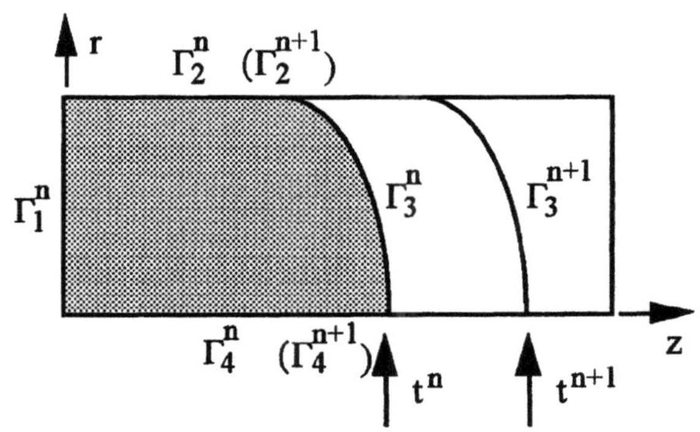

Figure 1. Flow domain and moving boundaries

For temperature boundary conditions in the non-isothermal context we adopt isothermal fixed temperatures on inlet and mould walls. On the symmetry boundary and flow front we assume an adiabatic condition. The thermal condition on the front is known to be somewhat more complicated than this and indeed at the next investigation stage we intend to apply a variable Robbins type condition of the form:

$$\frac{\partial T}{\partial s} = -Bi * T$$

where the Biot parameter Bi will vary by a factor from O(500) for air contact within the mould cavity to O(10) in the near mould wall vicinity.

For convenience we non-dimensionalise the above equations based on the experiments of Vos et al.[5], taking the radius of the cylinder at the inlet as the characteristic length L=0.008 m and the injection speed as the characteristic velocity V=0.0012 m sec^{-1}. The material parameters are the density ρ=1050 kg m^{-3}, the heat capacity c_p=1300 J kg^{-1} $^oK^{-1}$ and the thermal conductivity κ=0.15 w m^{-1} $^oK^{-1}$. For polystyrene data and the Carreau-Yasuda model, a reference viscosity μ_o=3873 Pa sec is associated with a zero shear rate and reference temperature T_o. Adopting a characteristic temperature scale of T=A_2 and assuming non-dimensionalised variables provides:

$$\mathbf{u}_t = \frac{1}{Re}\nabla.(\mu\nabla\mathbf{u}) - \mathbf{u}.\nabla\mathbf{u} - \nabla p$$

$$T_t = \frac{1}{Pé}\nabla^2 T - \mathbf{u}.\nabla T + \frac{G_f}{Pé}\mu\Phi$$

$$\mu = e^{(\frac{1}{T}-\frac{1}{T_o})}[1+(\lambda e^{\frac{A_1}{A_2}\frac{1}{T}}\gamma)^2]^{\frac{m-1}{2}}$$

where

$$Re = \frac{\rho LV}{\mu_o} = 2.5x10^{-6},\ Pé = \frac{\rho c_p LV}{\kappa} = 262,$$

$$G_f = \frac{\mu_o V^2}{\kappa A_2} = 3.3x10^{-6},\ \lambda = \frac{B_1 V}{L} = 10^{-11},$$

$$T_o = T_{melt} = 503^o K \text{ and } T_{wall} = 423^o K.$$

The momentum transport equation is dominated by viscous diffusion (coefficient $O(10^6)$), whilst the scales in the energy equation are

$$T_t = 3.8x10^{-3}\nabla^2 T - \mathbf{u}.\nabla T + 4x10^{-8}\mu\Phi.$$

The numerical scheme employed in this study is a so-called Taylor-Galerkin finite element method, which applies a temporal discretization in a Taylor series prior to the spatial Galerkin discretization. A pressure correction method is used within the present context of incompressible flow. A semi-implicit treatment for diffusion is adopted to address linear stability constraints. The structure of the scheme is one of some three distinct fractional stages. The first stage is a momentum-energy solution stage incorporating a predictor-corrector doublet. The second stage solves for a pressure increment on the time step and the last stage calculates a solenoidal velocity field at the end of the time step. Efficient solution is implemented by employing a direct Choleski method at stage two and a Jacobi iteration with mass lumping at all other stages. Further details on all these issues can be found in [6,7].

The Galerkin finite element spatial discretization adopted is continuous quadratic interpolation for velocity and temperature, and continuous linear interpolation for pressure over triangles in two dimensions (tetrahedra in three dimensions). The fully discretized equations for the semi-implicit Taylor-Galerkin/pressure correction scheme over a time step $\Delta t = t^{n+1} - t^n$ may be stated in matrix form as:

$$1a: [\frac{2}{\Delta t}\mathbf{M} + \frac{1}{2\mathrm{Re}}\mathbf{S}_u](\mathbf{U}^{n+\frac{1}{2}} - \mathbf{U}^n) = \frac{1}{2}(\mathbf{F}^n + \mathbf{F}^{n+\frac{1}{2}}) + \{-[\frac{1}{\mathrm{Re}}\mathbf{S}_u\mathbf{U} + \mathbf{N}(\mathbf{U})\mathbf{U}] + \mathbf{L}^T\mathbf{P}\}^n,$$

$$[\frac{2}{\Delta t}\mathbf{M} + \frac{1}{2\text{Pé}}\mathbf{S}_T](\mathbf{T}^{n+\frac{1}{2}} - \mathbf{T}^n) = \frac{1}{2}(\mathbf{F}^n + \mathbf{F}^{n+\frac{1}{2}}) \{-[\frac{1}{\text{Pé}}\mathbf{S}_T\mathbf{T} + \mathbf{N}(\mathbf{U})\mathbf{T}] + \frac{G_f}{\text{Pé}}\mu\Phi(\mathbf{U})\}^n,$$

$$1b: [\frac{1}{\Delta t}\mathbf{M} + \frac{1}{2\mathrm{Re}}\mathbf{S}_u](\mathbf{U}^* - \mathbf{U}^n) = \frac{1}{2}(\mathbf{F}^n + \mathbf{F}^{n+1}) + [-\frac{1}{\mathrm{Re}}\mathbf{S}_u\mathbf{U} + \mathbf{L}^T\mathbf{P}]^n - [\mathbf{N}(\mathbf{U})\mathbf{U}]^{n+\frac{1}{2}},$$

$$[\frac{1}{\Delta t}\mathbf{M} + \frac{1}{2\text{Pé}}\mathbf{S}_T](\mathbf{T}^{n+1} - \mathbf{T}^n) = \frac{1}{2}(\mathbf{F}^n + \mathbf{F}^{n+1}) - \frac{1}{\text{Pé}}\mathbf{S}_T\mathbf{T}^n + [-\mathbf{N}(\mathbf{U})\mathbf{T} + \frac{G_f}{\text{Pé}}\mu\Phi(\mathbf{U})]^{n+\frac{1}{2}},$$

$$2: \mathbf{K}(\mathbf{P}^{n+1} - \mathbf{P}^n) = -\frac{2}{\Delta t}\mathbf{L}\mathbf{U}^*,$$

$$3: \frac{1}{\Delta t}\mathbf{M}(\mathbf{U}^{n+1} - \mathbf{U}^*) = \frac{1}{2}\mathbf{L}^T(\mathbf{P}^{n+1} - \mathbf{P}^n),$$

where $\mathbf{U}^n$ is a nodal velocity vector at time t^n, $\mathbf{U}^*$ is an intermediate nonsolenoidal nodal velocity vector, $\mathbf{P}^n$ is a nodal pressure vector at time t^n, is a nodal pressure difference vector, **T** is a nodal temperature vector, **F** is a nodal forcing function vector due to boundary conditions, **K** is a pressure stiffness matrix, **L** is a divergence/pressure gradient matrix, **M** is a mass matrix, **N(U)** is an convection matrix, $\mathbf{S_u}$ is a momentum diffusion matrix, $\mathbf{S_T}$ is a thermal diffusion matrix, $\Phi(\mathbf{U})$ is a dissipation vector. The terminology employed here follows a previously established convention and a full explanation of all terms involved may be found in [10].

3. Front tracking methods

3.1 Compression-expansion scheme

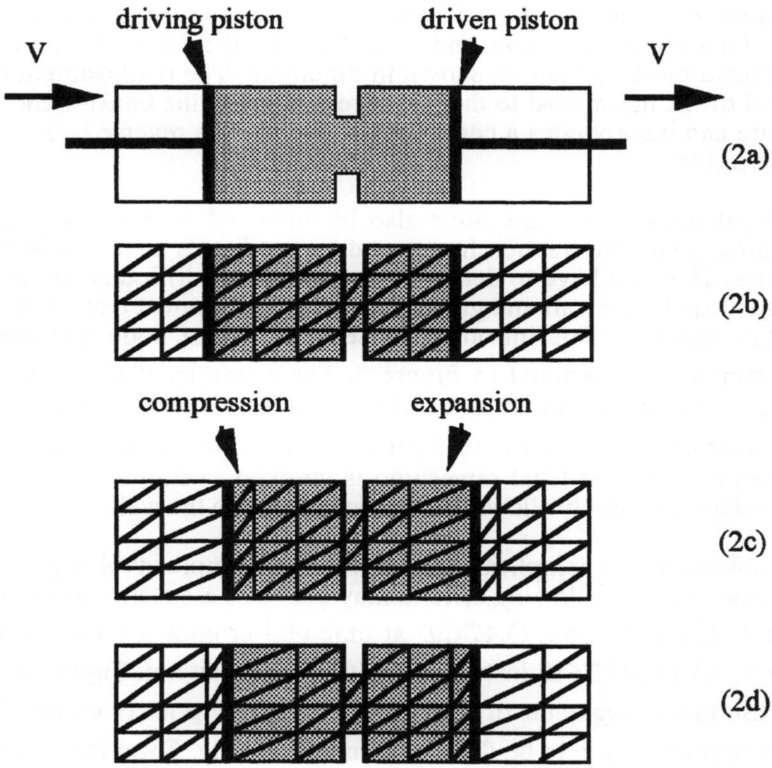

Figure 2. Illustration of compression-expansion scheme

We consider the application of this scheme for a piston-driven flow problem in a mould cavity which incorporates a contraction. This is a local mesh compression-expansion scheme based on particle tracking and is illustrated in Figure 2, where a shaded region indicates the filled wet part of the domain and an unshaded region indicates the dry part of the domain yet to be filled. Figure 2b illustrates a typical initial mesh applied to the domain, Figure 2c shows the situation after the pistons have moved, and Figure 2d exemplifies the mesh after local remeshing has been performed. The Taylor-Galerkin finite element analysis is only applied to the wet part of the domain. Both driving and driven pistons are considered to move across the mesh as the constriction is an immobile object. Particles are taken at the vertices of the mesh on the piston and tracked accordingly over each time step. This gives compression and expansion regions in the original mesh, being compressed in front of the driving piston and expanded behind the driven piston as shown in Figure 2c.

Certain limiting criteria are introduced to control the compression and expansion. For example at the driven piston, the mesh is updated when the front elements exceed one and a half times the size of an original background mesh element as shown in Figure 2d. The readjustment of the wet part of the mesh local to the piston reestablishes the background mesh structure and incorporates a perturbation upon it. The reverse is true at the driving piston.

This front tracking scheme must also be amended to take account of a deforming front with traction free boundary conditions as occurs in filling problems. Here we have to deal with the additional difficulty of particles moving close to the wall and giving rise to severe element distortion (long and thin regions). This introduces a second type of node and element readjustment as illustrated in Figure 3. For example, if the distance Δ between a particle B at the front and the wall is smaller than a prescribed small quantity δ as shown in Figure 3b, the flow front ABC will be switched to EFG and local remeshing is invoked as shown in Figure 3c. This strategy effectively allows the melt to move onto the wall.

For problems with a complicated geometry, for example involving a corner in the domain, the following criteria may prove of assistance as shown in Figure 4. If a flow front $O\mathcal{A}\mathcal{B}C\mathcal{D}$ at time t^{n+1} crosses a fixed reentrant corner P, take $O\mathcal{A}P\mathcal{B}C\mathcal{D}$ instead of $O\mathcal{A}\mathcal{B}C\mathcal{D}$ as shown in Figure 4a. This gives rise to two segments of the current front $O\mathcal{A}P$ prior to corner P, and $P\mathcal{B}C\mathcal{D}$ beyond P each to be dealt with independently. For a front segment PBC that has advanced some way further beyond the corner P, there are some additional practicalities to consider. There is now not only the proximity of particle B to the wall to adjust for, but also we must limit the extension of the front portion PB. As shown in Figure 4b if the distance

between particle B at the front PBC and the wall $\delta_z<\delta_{min}$ or $\delta_r>\delta_{max}$, take $\mathcal{A}$BC by internal extrapolation to the wall instead of PBC.

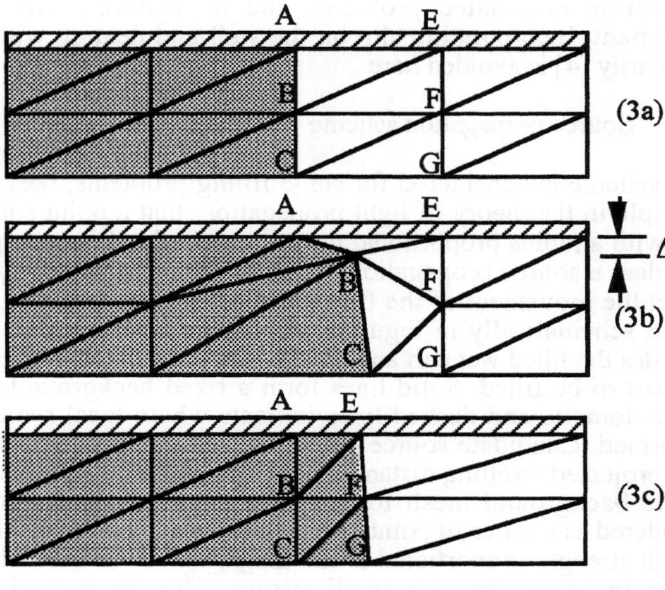

Figure 3. Front adjustment close to the wall

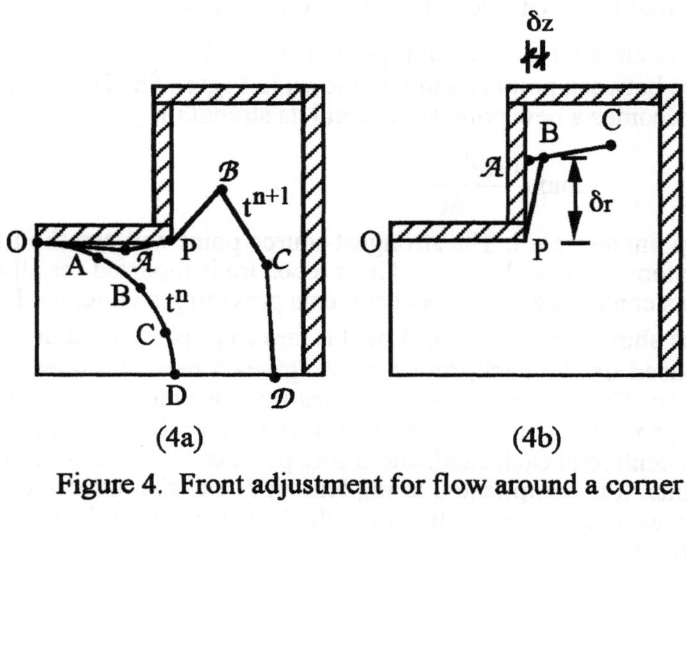

Figure 4. Front adjustment for flow around a corner

This scheme offers efficiency in that only strictly localised sections of the original mesh around the front are remeshed, avoiding the need to remesh the whole domain periodically. This method is appropriate for diffusion or convection dominated problems and is relatively straightforward to implement. Incorporating slip on the wall and dealing with contact point singularity [4] is avoided here.

3.2 Source propagation scheme

This scheme is considered for some filling problems, based on Huygens' principle in the theory of light propagation, that a point source can fill an area with a radius proportional to the magnitude of its strength. Following this idea, a source propagation technique was employed by Sitters[11] to model the movement of the flow front for filling problems, which can be shown schematically in Figure 5. As before a shaded part of the domain indicates the filled wet part and an unshaded part of the domain indicates the dry part to be filled. Solid lines form a fixed background mesh over the whole domain, and dashed lines indicate where local remeshing is to be performed. Candidate source points are sampled on the current front and their projected swelling distances are calculated with respect to intersection on the background mesh local to the front. A flow front is therefore considered as a series of point sources, that are indicated as circles in Figure 5, with strength proportional to the magnitude of its normal velocity $|\mathbf{u}|$ as shown in Figure 5a with small arrows. The dry part of the edge of a background element is divided into a number of edge points. An edge point j (indicated by •) can be filled by a point source i (shown as ∘) located in the same element as the edge point j if $(|\mathbf{u}_i|\Delta t-\delta_{ij}) > 0$, where δ_{ij} is the distance between points i and j as shown in Figure 5b. The filled edge point j then becomes a new point source and its strength is given by

$$\max\left(\frac{|u_i|\Delta t - \delta_{ij}}{\Delta t}\right)$$

for all point sources i. The strongest source point will wet the furthest edge points along a particular edge. This procedure is repeated for all background elements containing flow front segments providing an updated front position $O\mathcal{AB}$ as shown in Figure 5b. Local remeshing, performed around the flow front based on the background mesh, gives a new wet mesh as shown in Figure 5c. This scheme accommodates flow around sharp corners, but it is less appropriate for convection dominated problems, since the filling process centred at each candidate filling point is assumed to be uniform in all directions. It is complicated to implement due to the intricacies involved in local remeshing and projection onto the background mesh for each occasion a new front position is located.

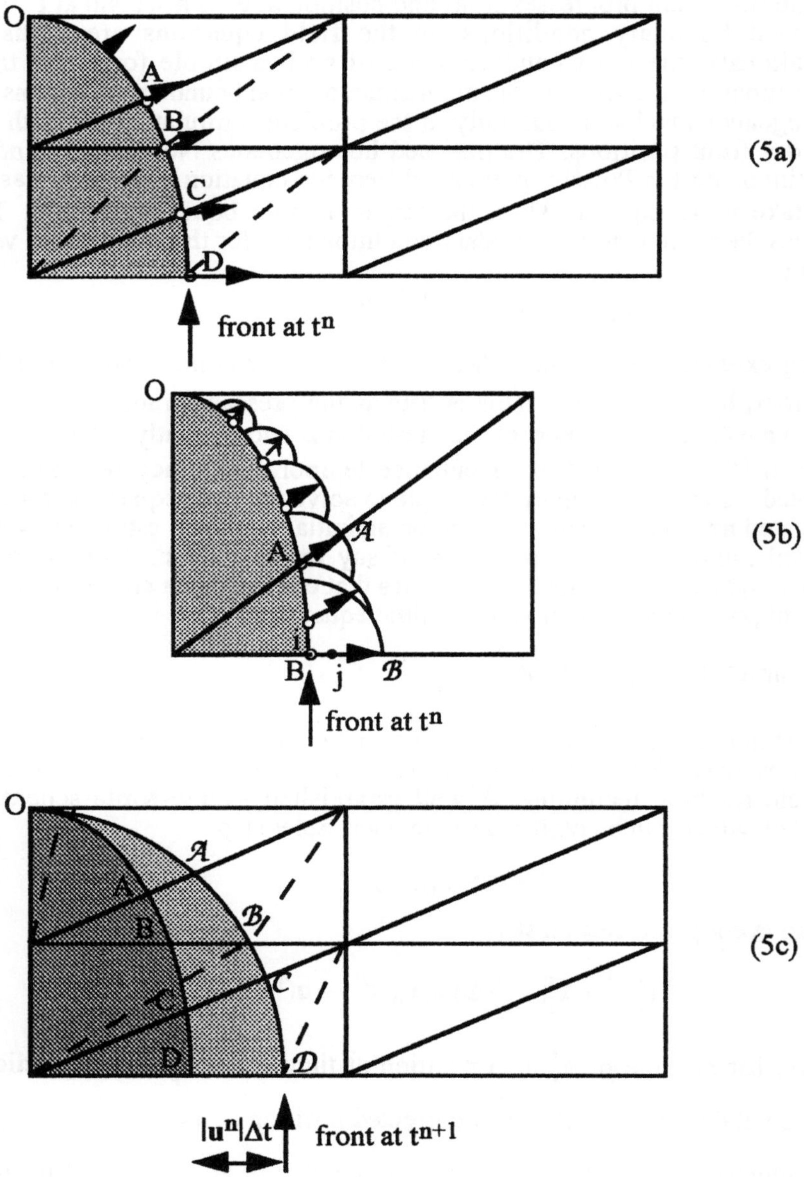

Figure 5. Illustration of source propagation scheme

With this updated wet mesh, the Taylor-Galerkin algorithm is invoked to evaluate the temporal increment over $[t^n, t^{n+1}]$ of the solution on the field.

As the flow front progresses from one position at t^n to a second at t^{n+1} the imposed boundary conditions on the field equations are adjusted. Fortuitously, the front conditions are of such a simple form that these alternations are ineffectual. The momentum traction boundary conditions are homogeneous and arise naturally in the problem formulation for both wet and dry front positions. The imposed homogeneous Neumann boundary conditions on the Poisson pressure difference equation still remain as we can take $\mathbf{u}^*$ to equal $\mathbf{u}^{n+1}$ on the larger domain boundary at t^{n+1}. This implies therefore that the boundary condition term for this equation in weak form is:

$$R = \oint_{\Gamma^{n+1}} \psi_i \nabla(p^{n+1} - p^n).\mathbf{s}d\Gamma$$

taking extended definition of ∇p^n over the wider domain with boundary specified by Γ^{n+1} at t^{n+1}. This is true as the pressure gradient term ∇p^n must vanish on the dry part of the mesh that is subsequently filled over the interval $[t^n, t^{n+1}]$. To further enhance temporal accuracy we have also adopted a sub-time stepping technique to solve the field equations for each estimated new front position. Based on an initial empirical estimation we set a fixed number of sub-time steps of say ten to fifteen. This approach accommodates for possible inaccuracies that can arise in a crude estimation of front position and solution of the fluid equations.

4. Particle tracking method

As it is important to track material deformation and particle histories behind the flow front during the filling process, it is necessary to incorporate a particle tracking technique. A well established Runge-Kutta scheme is adopted, which can be written as an explicit Euler step:

$${}^p\mathbf{x}_p^{n+1} = \mathbf{x}_p^n + \Delta t\mathbf{u}(\mathbf{x}_p, t^n)$$

and an averaged correction step:

$${}^c\mathbf{x}_p^{n+1} = \mathbf{x}_p^n + \frac{1}{2}\Delta t[\mathbf{u}(\mathbf{x}_p, t^n) + \mathbf{u}({}^p\mathbf{x}_p, t^{n+1})]$$

where, for particle p, $\mathbf{x}_p^n$ is a position at time t^n, ${}^p\mathbf{x}_p^{n+1}$ is a predicted position at time t^{n+1}, ${}^c\mathbf{x}_p^{n+1}$ is a corrected position at time t^{n+1}, $\mathbf{u}(\mathbf{x}_p, t^n)$ is a velocity at time t^n, $\mathbf{u}({}^p\mathbf{x}_p, t^{n+1})$ is a velocity at a predicted position ${}^p\mathbf{x}_p^{n+1}$ at time t^{n+1}. This scheme is performed after stage 3 of each time step of the Taylor-Galerkin algorithm as described in section 2. An alternative particle tracking scheme to consider is given by an explicit half step before stage 1a:

$$^{p}\mathbf{x}_p^{n+\frac{1}{2}} = \mathbf{x}_p^n + \frac{1}{2}\Delta t \mathbf{u}(\mathbf{x}_p, t^n),$$

and a correction step after stage 1a:

$$^{c}\mathbf{x}_p^{*} = \mathbf{x}_p^n + \Delta t \mathbf{u}(^{p}\mathbf{x}_p, t^{n+\frac{1}{2}}).$$

Further corrections are possible after stage 1b and the third stage. This scheme can be employed as a check on the accuracy of the above stated predictor-corrector scheme.

For filling problems, material bands are continuously traced from the inlet of the domain and different bands are traced at different times. For accuracy we have found it necessary to employ one hundred particles per unit length. Figure 6 shows this tracking procedure schematically. For practical convenience, a small gap is introduced between the first particle and the wall of the mould to avoid difficulties in dealing with fixed particles. To keep particles distributed uniformly along the band, it is wise to update the position of the neighbouring particles according to the average distance between any two particles. The position of particles along the band may be relocated if their neighbouring separation distances (δ) exceed certain limiting criteria, i.e., $\delta > \delta_{max}$ or $\delta < \delta_{min}$ with $\delta_{max} = 1.5\ \delta_{average}$ and $\delta_{min} = 0.5\ \delta_{average}$. This will involve a dynamic readjustment of the particle positions.

5. Numerical examples

5.1 Piston-driven flow with a contraction

This flow takes place in a cylinder, of unit radius and length 7.5 units, with a contraction ratio of 2:1 as shown in Figure 2a. The piston drives the flow at a constant speed, and since the constriction is a fixed obstacle, it is necessary here to solve the problem directly by moving the pistons across the mesh. Only half of the geometry is considered in the calculations and the front tracking is accomplished by the particle tracking method. A mesh with 416 triangles, 953 velocity nodes and 269 pressure nodes is employed yielding 2175 degrees of freedom as shown in Figure 7. An ABS polymer represented by a non-shifted power-law ($c \rightarrow 0$) is used for this problem under isothermal conditions, parameters are as shown in Table 1. Figure 7 shows the observed deformation patterns as the pistons move, and fountain flow is clearly observed. More elongation of the particle bands is observed in the core region of the cylinder than in the wall region because the axial velocity in the core region is larger than that in the wall region. The first

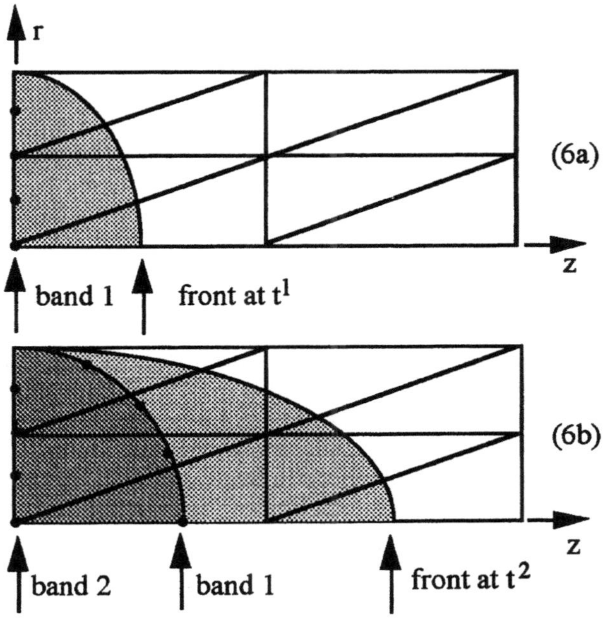

Figure 6. Particle tracking for material bands

Table 1.

	Polystyrene	ABS
m	0.48	0.52
A1 (°K)	11320	657
B1 (sec)	6.69×10^{-11}	0.0143
A2 (°K)	11285	
B2 (sec)	6.99×10^{-7}	

material band in front of the driving piston moves away from the wall and is gradually squeezed towards the core as the driving piston advances. This is due to the nature of the fountain flow, that provides an axial back flow in the boundary layer regions when viewed from a reference moving with the flow front. The back flow adjusts direction on meeting the driving piston and transfers towards the central core flow. The deformation pattern of all material bands is necked as they pass through the contraction. This is due to the contraction and expansion flows around the constriction, that introduce radial velocity components towards the centreline in the contraction flow and the reverse in expansion flow. These findings are in keeping with earlier evidence of deformation patterns for piston flow problems without a

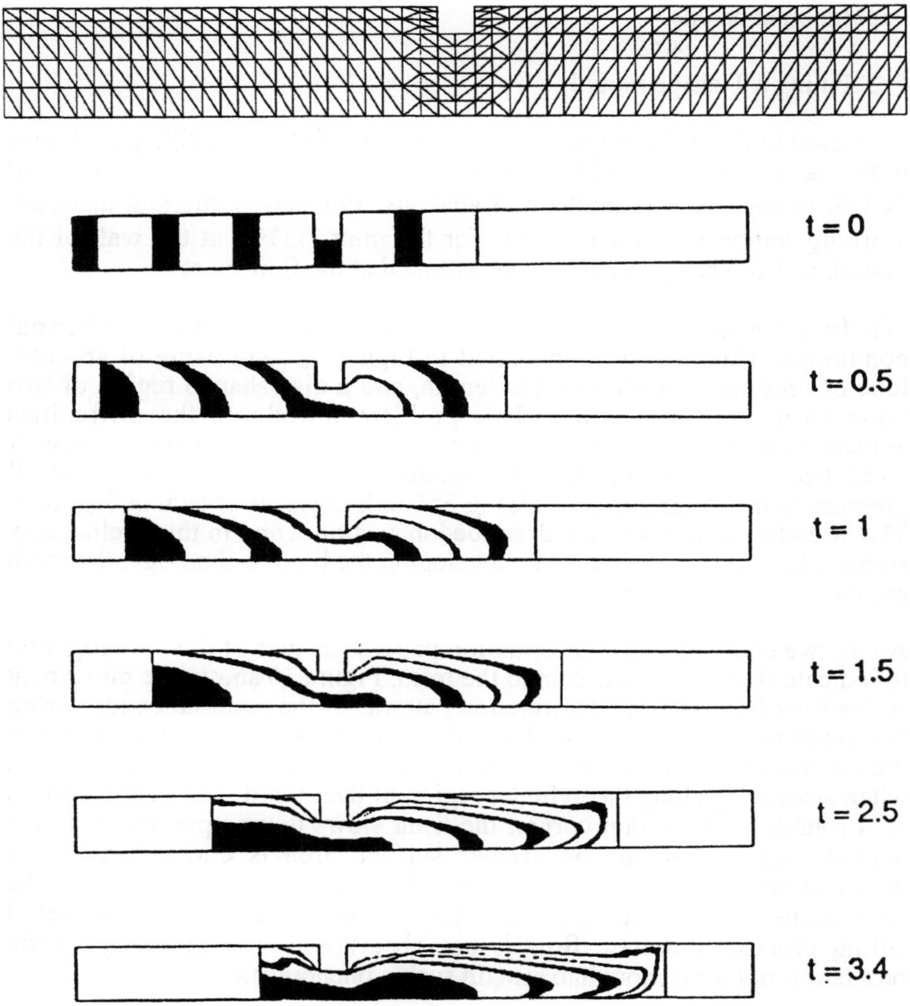

Figure 7. Deformation patterns of the moving piston problem

contraction [12,13]. Further investigation is underway for different contraction ratios and lengths, to compare with recently obtained experimental results.

5.2 Filling a T-shape mould

Data used to fit a polystyrene material is employed for both filling problems with material parameters as shown in Table 1, where a temperature of 503°K is used for the isothermal analysis. For non-isothermal analysis, working temperatures are 503°K for the melt, 423°K at the wall of the mould, and adiabatic conditions are assumed at the flow front.

We first consider the filling of a T-shaped mould under isothermal conditions. This mould is illustrated in Figure 8, comprising of an entry tube of unit radius and two units length, and a disc-shaped region of two units radius and unit width. This problem introduces the difficulties associated with tracking a moving front as it negotiates a reentrant corner. A fixed background mesh with 104 triangles, 243 velocity nodes and 70 pressure nodes is employed yielding 556 unknowns as shown in Figure 9. The boundary conditions are described in section 2 and in this preliminary study, where we are interested in developing the basic technology, the mesh employed is relatively coarse.

Again, we study the flow deformation patterns as part of this investigation to indicate flow behaviour behind the front. Figure 10 shows the movement of the flow front and the deformation patterns of the material bands during the filling process. The shape of the flow front is not particularly smooth as the mesh is relatively coarse. The bands deform initially in the axial direction when advancing along the cylinder region of the mould. The melt readjusts as it rounds the reentrant corner, the front slows down upon reaching the disc-shaped region and as greater surface area is encountered. The deformation patterns reorientate in the radial direction with the flow, and the front shape flattens. This gives a complete transient history of the overall filling process. Fountain flow is not observed as the geometry of the problem is not long enough to sustain such a phenomena.

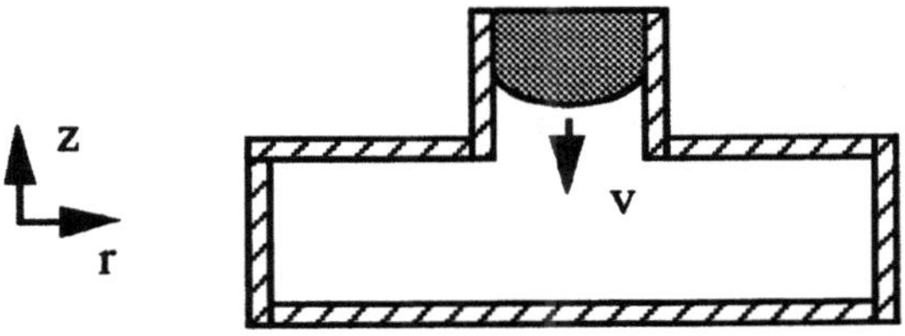

Figure 8. Schematic diagram for filling T-shaped mould

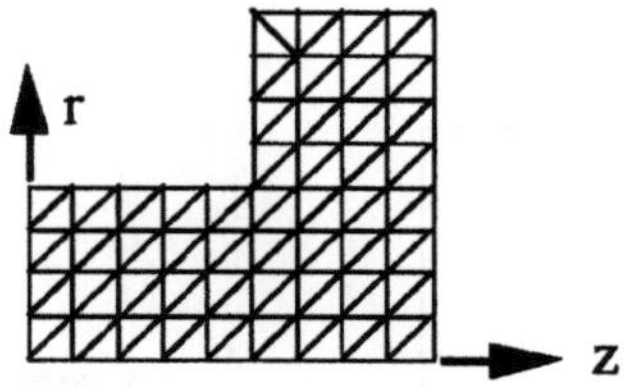

Figure 9. Mesh for T-shaped mould

5.3 Filling a cup-shaped mould

A second problem involves a cup-shaped mould as shown in Figure 11, and extends the geometrical aspects of the T-shaped filling problem to a cup-shape by addition of an annular region of four units outer radius, three units inner radius and three units in length beyond the disc. The T-shaped part of the domain is also now extended so that the disc radius is four times that of the down cylinder taken to the outer-most wall of the mould, the width remains at one unit. The simulation is considered both under isothermal and non-isothermal conditions. Two meshes are adopted as shown in Figure 12. Mesh 1 involves 288 triangles, 657 velocity/temperature nodes and 185 pressure nodes at the background mesh, giving 1499 degrees of freedom for the isothermal case and 2156 for the non-isothermal case. Mesh 2 is a hierarchical refinement of Mesh 1 and involves 1152 triangles, 2465 velocity/temperature nodes and 657 pressure nodes on the background mesh, giving 5587 degrees of freedom for the isothermal case and 9052 for the non-isothermal case. The boundary conditions are as taken in section 2.

Figures 13-14 show the movement of the flow front and the deformation patterns of the particle bands during the filling process under isothermal conditions on Mesh 1 and 2. The flow front negotiates not only the first reentrant corner of the mould smoothly but also the second. A smoother shape of the flow front is observed using the finer Mesh 2 compared with that from the coarser Mesh 1. This is because the accuracy of the source propagation scheme is dependent on the fineness of the background mesh. As with the T-shaped mould, a combination of different deformation patterns is observed in the various domain regions. In the disc-shaped region there is now a much larger radius for the melt to fill than in the previous problem, and here fountain flow can develop. This is observed, to start when the flow front approaches the second reentrant corner and becoming more prominent as the filling continues. It is noticeable that the velocity magnitude decreases as the cross-sectional area increases and with rounding the first reentrant corner, entering the disc-shaped region. The flow profile therefore flattens continuously as the melt travels along the disc and reaches a minimum state after rounding the second reentrant corner and entering the annular region. Subsequently the reduction in velocity magnitude is minimal and the profile adopts an almost flat plug flow state

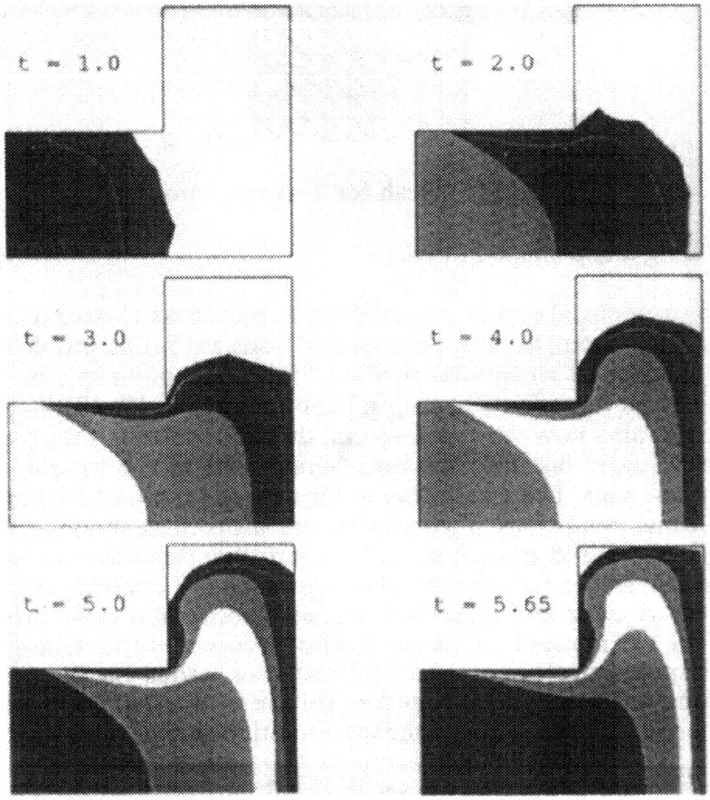

Figure 10. Material deformation patterns for T-shaped mould

thereafter.

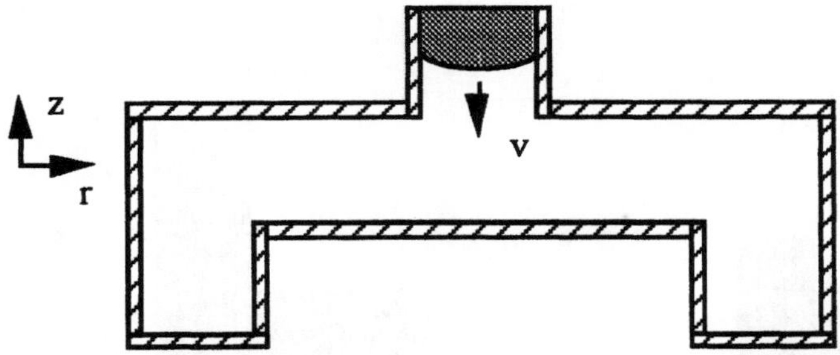

Figure 11. Schematic diagram for filling cup-shaped mould

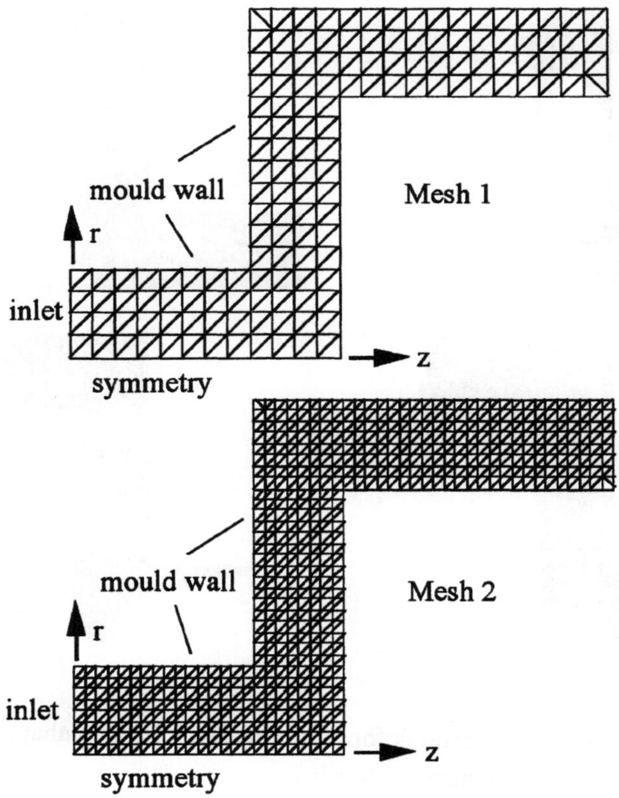

Figure 12. Meshes for cup-shaped mould

Non-isothermal conditions:

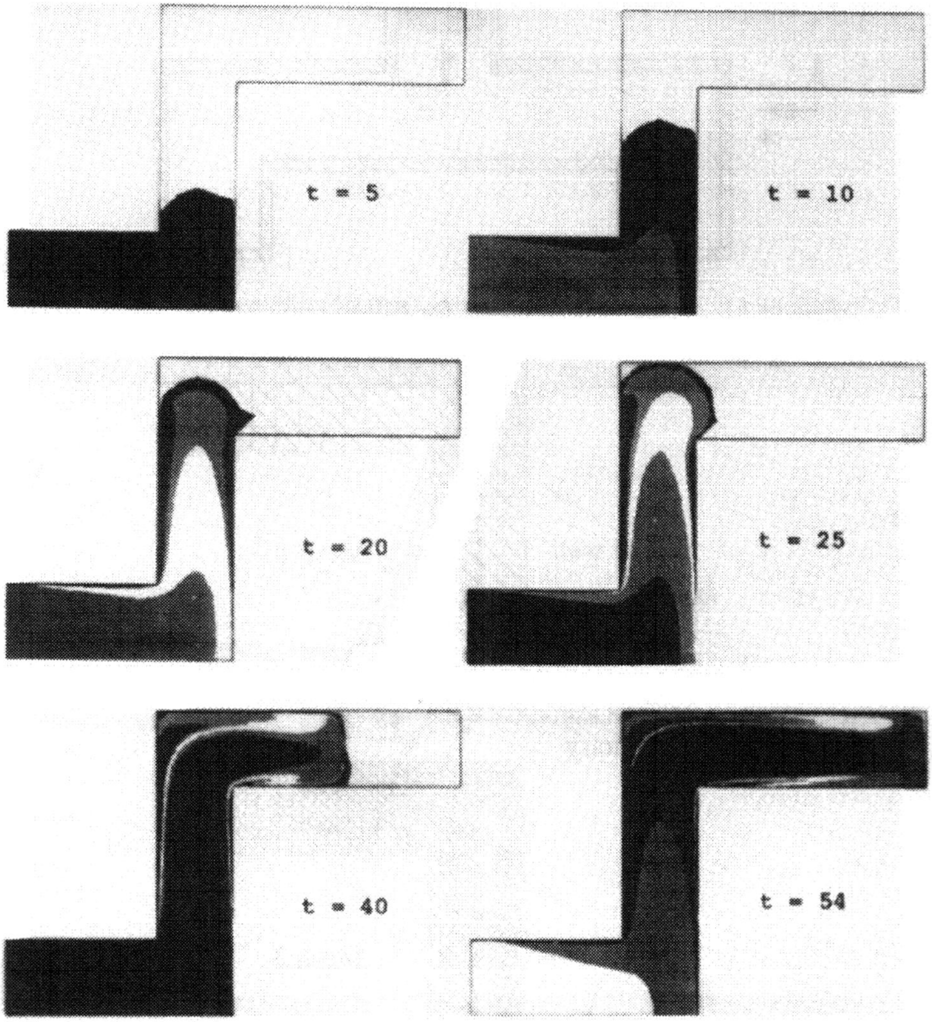

Figure 13. Material deformation patterns for cup-shaped mould; isothermal: Mesh 1

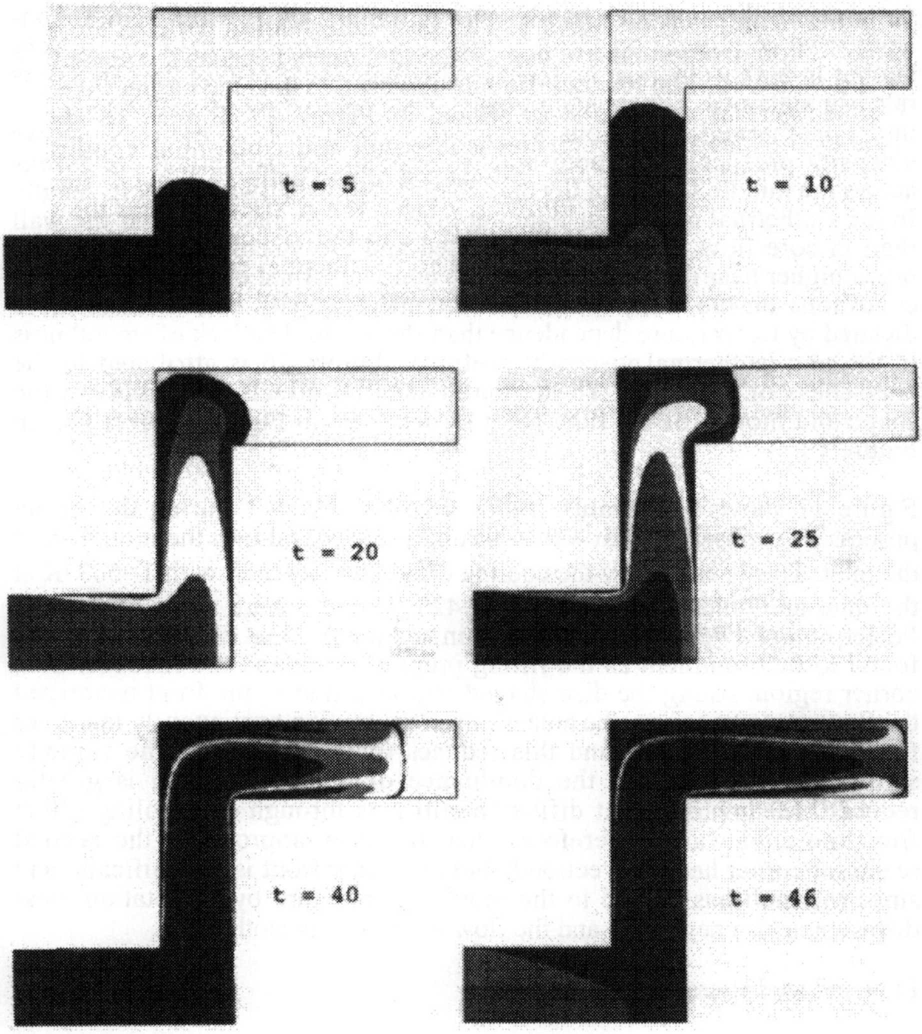

Figure 14. Material deformation patterns for cup-shaped mould; isothermal: Mesh 2

Figure 15 shows the movement of the flow front and the deformation patterns of the material bands during the filling process under non-isothermal conditions on Mesh 1. The flow deformation patterns are very similar in both isothermal and non-isothermal cases because the same flow rate is maintained. The fountain flow is observed to develop earlier than that under isothermal conditions as shown in Figure 13. Figure 16 shows viscosity profiles under both non-isothermal and isothermal conditions across the middle section of the disc-shaped region halfway along its length. In the isothermal case shear thinning gives a lower viscosity near the wall than in core flow, but this is readjusted and the viscosity becomes very much higher near the wall region in the non-isothermal case due to the cold conditions on the wall; the non-isothermal viscosity behaviour is more dictated by temperature dependence than shear rate. The lack of smoothness in the non-isothermal viscosity profile in Figure 16 is attributed to the coarseness of the mesh. These non-isothermal effects also improve the shape and motion of the flow front as observed in Figure 15 over that in Figure 13.

Figure 17 shows temperature fields given by Mesh 1 during the filling process. It is observed that heat is strongly convected into the mould from the inlet. This is caused by the cooling effect (the hot melt with $T=503^oK$ at the inlet and cold mould walls at $T=423^oK$) and a high value of thermal Péclet number Pé=262 as described in section 2. Heat convection is still found to be prominent as the filling progresses around the first reentrant corner region. Along the disc-shaped region and after the front negotiates the first reentrant corner, the area of the cross-section continuously increases for the same flow rate, and this reduces the magnitude of the velocity significantly. Therefore, the dominance of heat convection is greatly reduced. Meanwhile, heat diffusion still acts through the cooling effect from the cold walls. Therefore when the flow approaches the second reentrant corner heat convection behind the flow front is insignificant, and since the melt has cooled to the mould temperature by this station, heat diffusion is also negligible and the flow is practically isothermal.

Flow induced stress calculation:

Viscoelastic stresses occur in injected moulded articles through two separate phases, those induced dynamically with the flow and those that arise as a consequence of thermal cooling once injection has been completed. We are concerned here with only the flow induced stresses that are frozen into the moulded object in the injection phase of the process. By its nature this is a dynamic non-isothermal process, with stresses being built up and transmitted from one wet part of the spatial domain within the mould at time t^n to a second and new wet part of the domain at the next time step t^{n+1} as the flow front advances across the mould. Substantial changes in the stress

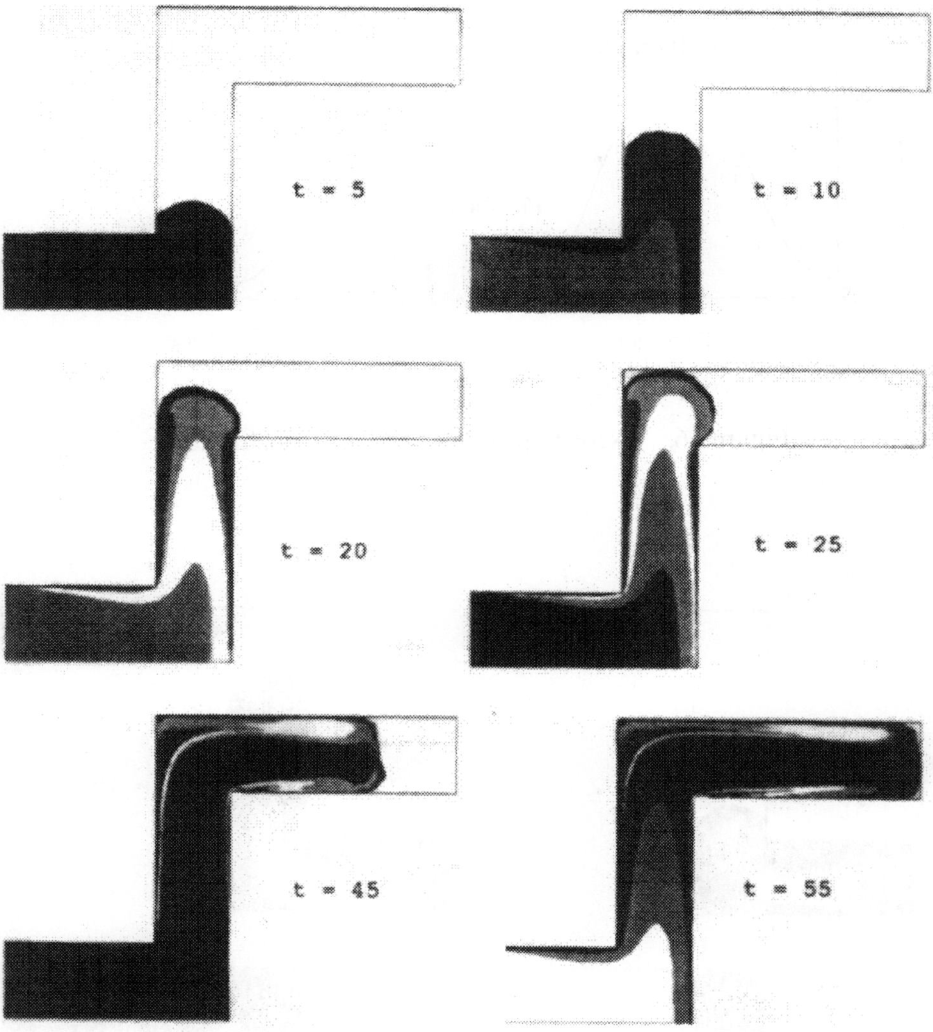

Figure 15. Material deformatio atterns for cup-shaped mould;
non-isothermal: Me

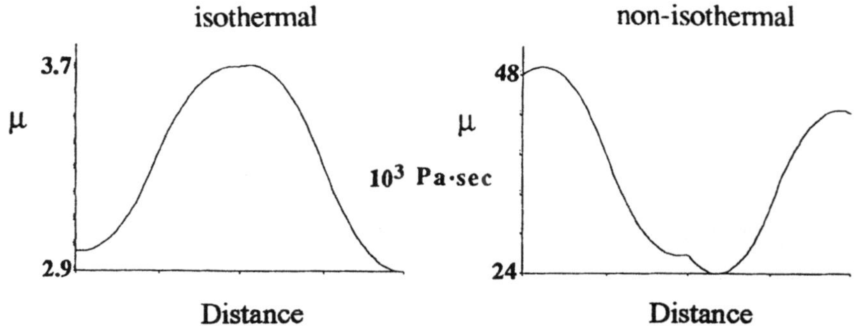

Figure 16. Viscosity profiles, t=20, disc position r=2

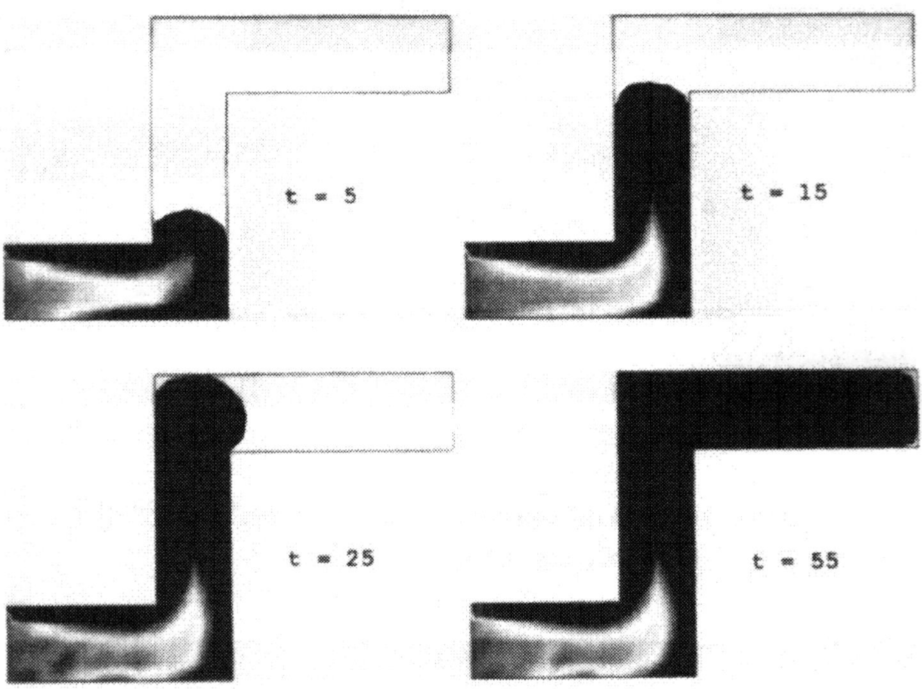

Figure 17. Temperature contours; non-isothermal: Mesh 1

may be generated in this manner, particularly so as the processing conditions are non-isothermal.

We may adopt here differential constitutive models to represent the viscoelastic stresses and thus far we have tested Maxwell/Oldroyd and Phan-Thien/Tanner models[14]. For constitutive models under non-isothermal conditions see [15], where relaxation time and viscosity have some temperature dependence. The energy equation will be taken as stated above with a purely viscous heat dissipation function as a first approximation.

The intention is to solve for the viscoelastic stresses separately from the velocity-temperature-pressure field equations on each shift of the spatial domain over each time interval $[t^n, t^{n+1}]$. That is whilst solving for the stresses, the velocity temperature field data will remain frozen. This decoupled approach has worked effectively in other instances on fixed domains[16]. This may be implemented with or without extra-stress feedback into the momentum equations. If no stress feedback is permitted then the approach will be fully viscous. Nevertheless since the injection phase is largely a shear dominated flow this approach is anticipated to provide adequate results. At a later investigation stage it is quite possible to solve for both stress and other primary variables together within the context of our time-stepping scheme, at which point feedback will be incorporated and the accuracy of the above premise can be tested.

6. Conclusions

We have been able to display some plausible results for axisymmetric filling problems, testing the viability of two different front tracking schemes and a particle tracking method. Three dimensional analysis is now underway for which the compression-expansion technique is preferred in the first instance. Both isothermal and non-isothermal processing conditions have been implemented and the Taylor-Galerkin code has proved an effective vehicle to achieve this. Complex three-dimensional moulds and a flow induced residual stress analysis await further investigation.

Acknowledgement

Financial support from the Commission of the European Communities is gratefully acknowledged (Contract BREU-495-BRITE/EURAM, Project BE-4076). Many thanks to Mr T Simpson for his help in flow visualisation.

References

1. Rose, W.: 1961, 'Fluid-fluid Interfaces in Steady Motion', *Nature*, 191 (4785), 242-243.

2. Tadmor, Z.:1974, 'Molecular Orientation in Injection Molding', *J. Appl. Polym. Sci.*, 18, 1755.
3. Coyle, D.J., Blake, J.W. and Macosko, C.W.: 1987, 'The Kinematics of Fountain Flow in Molding Filling', *AIChE J.*, 33 (7), 1168-1177.
4. Mavridis, H., Hrymak, A. N. and Valachopoulos, J.: 1988, 'Transient Free-surface Flows in Injection Mold Filling', *AIChE J.*, 34 (3), 403-410.
5. Vos, E., Meijer, H. E. H. and Peters, G. W. M.: 1991, 'Multilayer Injection Molding', *Int. Polym. Processing*, VI, 42-50.
6. Townsend P. and Webster, M.F.: 1987, 'An Algorithm for the Three-dimensional Transient Simulation of Non-Newtonian Fluid Flow', *Transient/Dynamic Analysis and Constitutive Laws for Engineering Materials*, Eds. Pandé, G.N. and Middleton, J., Nijhoff, T12/1.
7. Hawken, D.M., Tamaddon-Jahromi, H.R., Townsend P. and Webster, M.F.: 1990,' A Taylor-Galerkin Based Algorithm for Viscous Incompressible Flow', *Int. J. Num. Meth. Fluids*, 10 (3), 327-351.
8. Hassager, O., Henriksen, P., Townsend, P., Webster, M.F. and Ding, D.: 1991 'The Quarterbend: A Three-dimensional Benchmark Problem', *Computers Fluids*, 20 (4),373-386.
9. Ding, D., Townsend P. and Webster, M.F.: 1993, 'Transient Simulation of Non-Newtonian Non-isothermal Flows in Two and Three-dimensions', *Proc. Numerical Methods for Thermal Flow*, ed. R.W. Lewis, Pineridge Press, VIII (I) 621-633,
10. Ding, D., Townsend P. and Webster, M.F.: 1993, 'Computer Modelling of Transient Thermal Flows of Non-Newtonian Fluids', *Int. J. Non-Newt. Fluid Mech.*, 47, 239-65.
11. Sitters, C.W.M.: 1988, 'Numerical Simulation of Injection Moulding', *PhD thesis*, Eindhoven.
12. Ding, D., Townsend, P. and Webster, M.F.: 1992, 'Some Studies of Non-Newtonian Non-isothermal Flows in Multi-layer Injection Moulding', *Proc. Numerical Methods in Industrial Forming Processes-NUMIFORM 92*, ed. J.-L. Chenot, R.D. Wood & O.C. Zienkiewicz, Balkema, 351-356.
13. Townsend, P. and Webster, M.F.: 1992, 'Problems Associated with the Simulation of Multi-layer Injection Moulding', *Proc. Ist International Conference on Transport Phenomena*, Elsevier, 1099-1113.
14. Carew, E.O., Townsend, P. and Webster, M.F.: 1993, 'A Taylor-Petrov-Galerkin Algorithm for Viscoelastic Flow', *Int. J. Non-Newt. Fluid Mech.*, 50, 253-287.
15. Crochet , M.J., Debae, F. and Goublomme, A.: 1992, 'Recent Progress on Viscoelastic Extrusion', *Proc. Numerical Methods in Industrial Forming Processes- NUMIFORM 92*, ed. J.-L. Chenot, R.D. Wood & O.C. Zienkiewicz, Balkema, 15-23.
16. Carew, E.O., Townsend, P. and Webster, M.F.: 1993, 'Taylor-Galerkin Algorithms for Viscoelastic Flow-Application to a Model Problem; in press *Int. J. Num. Meth. PDE.*

NUMERICAL MODELLING OF NONISOTHERMAL VISCOELASTIC FLOW BETWEEN ECCENTRICALLY ROTATING CYLINDERS

X.K.LI and A.R.DAVIES
Department of Mathematics
University of Wales
Aberystwyth SY23 3BZ
United Kingdom

Abstract. Two-dimensional nonisothermal viscoelastic flow between eccentrically rotating cylinders is investigated numerically for a range of thermal boundary conditions and different values of eccentricity. The constitutive model employed is a White-Metzner fluid, with a viscosity and single relaxation time both of which are shear thinning, temperature thinning and pressure thickening. A pseudospectral method with bipolar transformation is used for spatial discretization of the governing equations, and a first order time stepping scheme with a filtered pressure step is used for time integration within the constraint of incompressibility.

It is shown that a Dirichlet condition for temperature on the inner cylinder, coupled with a Biot (or Robbins) condition on the outer cylinder leads to equilibrium temperature fields which have maxima in the small gap, as may be expected in journal bearing lubrication. The same effects are achieved with Biot conditions on both cylinders. The results are sensitive to choice of Biot numbers. It is also shown that at high eccentricity the relaxation time can increase by at least one order of magnitude. At very high eccentricity there is therefore an indication that an increase in load-bearing capacity may be possible due to the effect of viscoelastic normal stresses. The pressure fields at high eccentricity are also of interest since they display a breakdown of antisymmetry, with a trend towards higher positive pressures. This lends weight to the suggestion that a viscoelastic fluid is less likely to cavitate than a purely viscous fluid.

Key words: eccentric cylinder, nonisothermal viscoelastic flow, Biot number, journal bearing lubrication, pseudospectral method

1. Introduction

The two-dimensional flow of a viscoelastic fluid in a journal bearing under static loading conditions is an interesting prototype problem for several reasons. First, significant stress boundary layers can develop in the flow at high eccentricity, even at modest rotational speeds of the inner cylinder and at modest relaxation times of the fluid [3]. The accurate resolution of such boundary layers offers an important challenge to modern numerical methods. In particular, the very high number of degrees of freedom demanded by traditional finite-element discretization in attempting to resolve these layers [4] has led to other approaches being tried, namely high order methods such as spectral and spectral element methods. The smoothness and simplicity of the geometry is of advantage in this context, and some success has already been reported for isothermal flows [3, 9].

Secondly, there is the practically important question of what are the beneficial effects of viscoelasticity (if any) in journal bearing lubrication. For example, can nor-

J.F. Dijksman and G.D.C. Kuiken (eds.),
IUTAM Symposium on Numerical Simulation of Non-isothermal Flow of Viscoelastic Liquids, 159-177.

mal stresses induced by viscoelasticity compensate for the adverse effects of shear thinning, namely higher wear resulting from lower friction. The published evidence in favour of positive benefits is not conclusive (see, for example, [2, 12]). In connection with the influence of viscoelasticity on load-bearing capacity, however, the constant viscosity, isothermal study reported in [9] reveals that relaxation times of order 10^{-4}s are required before viscoelasticity increases effective load-bearing capacity. In lubricating oils under isothermal, isobaric conditions, realistic relaxation times are of order 10^{-6}s, well below values of any significance.

The role of viscoelasticity in journal bearing lubrication cannot be isolated from a third factor of interest, namely the effects of temperature and pressure. Viscosity increases under pressure but decreases as the temperature increases. Relaxation times also behave in similar fashion in responding to temperature and pressure. Two questions of relevance here are: can significant increases in normal stress be found at high pressures, and what is the effect of temperature on these stress levels. Before numerical simulation can answer these questions, however, we must know what are the correct boundary conditions to use for temperature, at least when we treat the journal bearing in isolation from its immediate mechanical configuration. This is the topic addressed in this paper. In particular we investigate the questions of what temperature boundary conditions appear to be realistic, and how sensitive are flow simulations to changes in temperature boundary conditions.

2. Mathematical modelling

2.1. The geometry

Consider the two-dimensional geometry shown schematically in Fig. 1. The journal of radius R_J rotates with angular velocity Ω in a stationary bearing of radius R_B. Both the journal and the bearing are assumed to be of infinite extent in the axial z-direction. The axes of the journal and the bearing are separated by a distance e, and it is customary to define an eccentricity

$$\epsilon = \frac{e}{c}, \tag{1}$$

where $c = R_B - R_J$, so that $0 \leq \epsilon \leq 1$.

2.2. The governing equations

One of the simplest viscoelastic fluid models available, which allows for variable viscosity and a variable relaxation time (one mode), is the White-Metzner model [13], which reduces to the Upper Convected Maxwell (UCM) model when the viscosity and relaxation time are kept constant. In this preliminary study, therefore, we restrict attention to the White-Metzner constitutive model. Nonisothermal incompressible unsteady flow between the cylinders is then governed by a system of seven partial differential equations of mixed type, consisting of:
two momentum equations

$$\rho \left(\frac{\partial \mathbf{v}}{\partial t} + \mathbf{v} \cdot \nabla \mathbf{v} \right) = -\nabla p + \nabla \cdot \mathbf{T}; \tag{2}$$

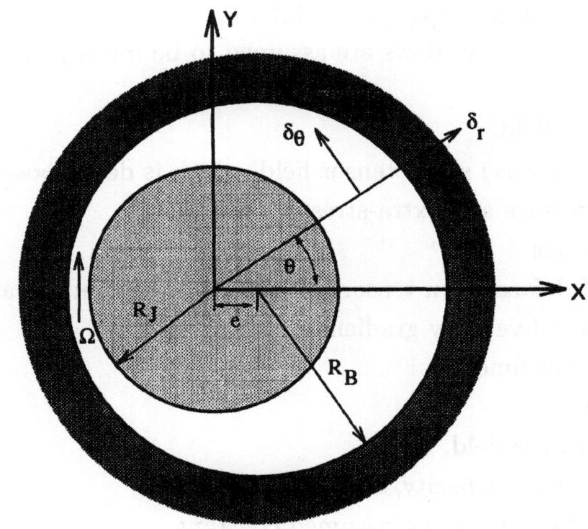

Fig. 1. Geometry of eccentric cylinder model

the scalar conservation of mass equation

$$\nabla \cdot \mathbf{v} = 0; \tag{3}$$

three constitutive equations

$$\mathbf{T} + \lambda \overset{\nabla}{\mathbf{T}} = 2\eta \mathbf{d}; \tag{4}$$

where

$$\mathbf{d} = \frac{1}{2}\left(\nabla \mathbf{v} + (\nabla \mathbf{v})^T\right), \tag{5}$$

$$\overset{\nabla}{\mathbf{T}} = \frac{\partial \mathbf{T}}{\partial t} + \mathbf{v} \cdot \nabla \mathbf{T} - (\nabla \mathbf{v}) \cdot \mathbf{T} - \mathbf{T} \cdot (\nabla \mathbf{v})^T; \tag{6}$$

and the scalar energy equation

$$\rho C_p \left(\frac{\partial \Theta}{\partial t} + \mathbf{v} \cdot \nabla \Theta\right) = \nabla \cdot (\kappa \nabla \Theta) + (\boldsymbol{\sigma} : \nabla \mathbf{v}), \tag{7}$$

where

$$\boldsymbol{\sigma} = -p\mathbf{I} + \mathbf{T}, \tag{8}$$

and

$$(\boldsymbol{\sigma} : \nabla \mathbf{v}) = \text{trace}(\boldsymbol{\sigma} \nabla \mathbf{v}). \tag{9}$$

The notation in equations (2)-(7) is as follows:

- ρ – density. Since all flows are assumed to be incompressible, ρ is a constant.
- $\mathbf{v}$ – velocity field.
- p – pressure field.
- σ – total (Cauchy) stress tensor field, which is decomposed in (8) into pressure and extra-stress.
- $\mathbf{I}$ – unit tensor.
- $\mathbf{d}$ – rate of deformation tensor, which is the symmetric part of the spatial velocity gradient.
- λ – relaxation time.
- η – viscosity.
- Θ – temperature field.
- C_p – specific heat capacity, assumed constant.
- κ – thermal conductivity, assumed constant.
- $\overset{\nabla}{\mathbf{T}}$ – the upper convected derivative of the extra-stress $\mathbf{T}$.

The final term in the energy equation (7) refers only to viscous heating [14], and so there is no implicit modelling of elastic energy effects. Equation (7), therefore, is almost certainly not the most general form of the energy equation describing non-isothermal viscoelastic behaviour. The present authors, however, have been unable to find any adequate treatment in the literature of the most appropriate energy equation to use in this context. We therefore adopt equation (7) as a temporary measure, with a view to using a better model when available.

2.3. Viscosity and relaxation time

The White-Metzner constitutive relations (4) are incomplete without modelling the dependence of viscosity and relaxation time on shear-rate, strain-rate, temperature and pressure. There is little option here but to adopt an heuristic approach. For viscosity, therefore, we propose the following model:

$$\eta = \{\eta_\infty + (\eta_0 - \eta_\infty)[1 + (K_s\dot{\gamma})^m]^{-1}[1 + (K_e\dot{\epsilon})^n]^{-1}\}$$

$$\times \exp\left(\tfrac{\alpha}{3}\,\mathrm{trace}(\sigma) + \tfrac{\beta}{\Theta}\right), \tag{10}$$

where the shear-rate $\dot{\gamma}$ and strain-rate $\dot{\epsilon}$ are defined, respectively, by

$$\dot{\gamma} = \sqrt{2\mathrm{trace}\,\mathbf{d}^2},$$

and

$$\dot{\epsilon} = \frac{6\det\mathbf{d}}{\mathrm{trace}\,\mathbf{d}^2},$$

and the coefficients K_s and K_e, although independent of $\dot{\gamma}$ and $\dot{\epsilon}$, are dependent on temperature and pressure. In particular, following [7] we choose

$$K_s = K_s(\Theta, p) = \exp\left(\frac{\bar{\alpha}}{3}\text{trace}(\boldsymbol{\sigma}) + \frac{\bar{\beta}}{\Theta} + E\right). \tag{11}$$

The constants $\eta_0, \eta_\infty, m, n, \alpha, \bar{\alpha}, \beta, \bar{\beta}$ and E are material parameters which must be estimated by best-fitting available experimental data.

The model described by equation (10) imparts both shear thinning and extension thinning behaviour to the viscosity, according to separate formulae of Cross type [1]. Pressure thickening and temperature thinning are modelled separately through simple exponential laws [1]. We note that in equation (10) the effective pressure is taken to be $\frac{1}{3}\text{trace}(\boldsymbol{\sigma})$, which reduces to the actual pressure in the case of a Newtonian fluid. Two sets of exponential parameters are used. The first set (α, β) in equation (10) controls the dependence of the low shear-rate and low strain-rate plateau on temperature and pressure. The dependence of the high shear-rate and high strain-rate plateau on temperature and pressure is assumed to be governed by the same exponential law. The second set of parameters $(\bar{\alpha}, \bar{\beta})$ control the dependence of the rate of shear thinning on temperature and pressure, which may realistically assumed to be independent of the low and high plateaux levels reached. In principle, one could attempt to control the dependence of the rate of extension thinning on temperature and pressure in a similar way using additional parameters, but we make no such attempt in this paper.

Finally, we must specify the dependence of relaxation time λ on shear-rate, strain-rate, temperature and pressure. Whatever model we choose in this context must be consistent with known viscometric properties of lubricating oils. Independently of temperature and pressure, under viscometric conditions the first normal stress-difference N_1 in lubricating oils behaves empirically as [1]

$$N_1 = 2k_t(\eta\dot{\gamma})^q, \tag{12}$$

where k_t and q are constants, $q \approx 2$. In addition, it can be shown that the White-Metzner model under viscometric conditions yields

$$N_1 = 2\eta\lambda\dot{\gamma}^2. \tag{13}$$

Comparing equations (12) and (13) when $q = 2$ we find

$$\lambda = k_t\eta. \tag{14}$$

We shall adopt (14) as the simplest model for relaxation time, on the understanding that η behaves according to equation (10).

3. Boundary conditions

We assume no-slip velocity boundary conditions on the journal and the bearing at all times. Furthermore, since there are no inlet boundaries in the geometry, stress boundary conditions are not required. It is sufficient to specify an initial state of rest at time $t = 0$. In this case, however, care must be taken to avoid an impulsive start to the rotation of the journal, which can result in a large jump in pressure with an unrealistic increase in viscosity. The rotational speed of the journal should be increased smoothly from rest. In this paper we shall be mainly concerned with the boundary conditions for temperature.

3.1. Thermal boundary conditions

Let Γ denote either the journal boundary or the bearing boundary. We examine combinations of the following two types of condition:
a Dirichlet condition

$$\Theta = \Theta_\Gamma, \qquad \text{on} \quad \Gamma; \tag{15}$$

and a Biot or Robbins condition

$$\frac{\partial \Theta}{\partial n}\Big|_\Gamma = \frac{Bi}{h}(\Theta_r - \Theta_\Gamma), \qquad \text{on } \Gamma. \tag{16}$$

Here Bi denotes a Biot number, h a characteristic thickness, Θ_r a reference temperature, and Θ_Γ the temperature on the boundary Γ. Condition (15) can be viewed as the limiting case of (16) with $Bi \to \infty$. The other extreme, $Bi = 0$, gives rise to the adiabatic boundary condition

$$\frac{\partial \Theta}{\partial n} = 0, \qquad \text{on } \Gamma, \tag{17}$$

favoured by some authors [6, 8]. We shall consider mostly positive values of Bi in this paper.

We consider the following four combinations:

DJ+DB

$$\begin{aligned} &\Theta = \Theta_J, && \text{on journal;} \\ &\Theta = \Theta_B, && \text{on bearing.} \end{aligned} \tag{18}$$

DJ+RB

$$\begin{aligned} &\Theta = \Theta_J, && \text{on journal;} \\ &\frac{\partial \Theta}{\partial n} = \frac{Bi}{h}(\Theta_{rB} - \Theta_B), && \text{on bearing.} \end{aligned} \tag{19}$$

RJ+DB

$$\begin{aligned} &\frac{\partial \Theta}{\partial n} = \frac{Bi}{h}(\Theta_{rJ} - \Theta_J), && \text{on journal;} \\ &\Theta = \Theta_B, && \text{on bearing.} \end{aligned} \tag{20}$$

RJ+RB

$$\begin{aligned} &\frac{\partial \Theta}{\partial n} = \frac{Bi}{h}(\Theta_{rJ} - \Theta_J), && \text{on journal;} \\ &\frac{\partial \Theta}{\partial n} = \frac{Bi}{h}(\Theta_{rB} - \Theta_B), && \text{on bearing.} \end{aligned} \tag{21}$$

In case (21) the values of Bi on the journal and bearing are taken to be equal.

3.2. ESTIMATING THE BIOT NUMBER

The value of Bi on the bearing can be estimated very easily from a consideration of heat balance. Heat flux on the bearing wall is given by

$$q_B = -\kappa_f \left(\frac{\partial \Theta}{\partial n} \right) |_B = -\kappa_f \frac{Bi}{h} (\Theta_r - \Theta_B), \tag{22}$$

where κ_f denotes the thermal conductivity of the fluid. On the other hand the heat flux through the wall of a cylindrical pipe of inner radius R_B, wall thickness s, thermal conductivity κ_B, and inner and outer temperature Θ_B and Θ_r, may be shown to be

$$q_B = \frac{\kappa_B}{R_B} \frac{(\Theta_B - \Theta_r)}{\log(1 + s/R_B)}. \tag{23}$$

Taking the characteristic length h in (22) to be the radial difference $R_B - R_J$, it follows from (22) and (23) that on the bearing

$$Bi = \frac{\kappa_B}{\kappa_f} \frac{R_B - R_J}{R_B \log(1 + s/R_B)}. \tag{24}$$

Estimating Bi on the journal is not so easy. Repetition of the above argument leads to

$$Bi = \frac{\kappa_J}{\kappa_f} \frac{R_B - R_J}{R_J \log(R_J/R)}, \tag{25}$$

where R is some *chosen* radius within the journal. The choice of $R = 0$ leads to the adiabatic condition (16), whereas the choice $R = R_J$ leads to the Dirichlet condition (15).

4. Numerical method

Equations (2)-(7) are solved using a pseudospectral method for spatial discretization and a first-order time-stepping method for time integration. The incompressibility condition (3) is imposed at each time step using an intermediate projection step for pressure, which is a slight modification of Chorin's scheme [5].

First the eccentric annulus in Fig. 1 is mapped onto a concentric annulus

$$\{(\xi, \zeta)) : \ \xi_J \leq \xi \leq \xi_B, \, 0 \leq \zeta \leq 2\pi\},$$

using the bipolar transformation

$$x = \frac{a \sinh \xi}{\cosh \xi + \cos \zeta}, \qquad y = \frac{a \sin \zeta}{\cosh \xi + \cos \zeta}, \tag{26}$$

where

$$a = \frac{1}{2e} \sqrt{(R_J^2 + R_B^2 - e^2)^2 - 4R_J^2 R_B^2}, \tag{27}$$

There are seven dependent variables in equations (2)-(7), namely the two components of velocity u, v; the three components of extra-stress $T^{\xi\xi}, T^{\xi\zeta}$ and $T^{\zeta\zeta}$; the pressure p and temperature Θ. We calculate nodal values for each of the seven variables on a Chebyshev-Fourier grid on the concentric annulus, defined by

$$\left\{\left(\xi_j = \cos\left(\frac{j\pi}{M}\right), \zeta_k = \frac{2k\pi}{N+1}\right);\ 0 \le j \le M,\ 0 \le k \le N\right\}, \tag{28}$$

where M and N are positive integers determining the order of the trial functions in the radial and azimuthal directions, respectively. Each variable has a continuous pseudospectral representation of the form

$$\phi(\xi,\zeta) = \sum_{j=0}^{M}\sum_{k=0}^{N} \phi_{jk} g_j(\xi) h_k(\zeta), \tag{29}$$

where the $g_j(\xi)$ are the Lagrange interpolating polynomials (of degree M) for the radial Chebyshev grid and the $h_k(\zeta)$ are the interpolating trigonometric polynomials for the uniform Fourier grid in the azimuthal direction. The values ϕ_{jk} in (29) are the nodal values $\phi(\xi_j, \zeta_k)$.

The nodal values of the spatial derivatives of ϕ are then related to the nodal values ϕ_{jk} by differentiating (29):

$$\begin{aligned}\frac{\partial\phi}{\partial\xi}(\xi_j,\zeta_k) &= \sum_{m=0}^{M}\sum_{n=0}^{N} \phi_{mn} g'_m(\xi_j) h_n(\zeta_k),\\ \frac{\partial\phi}{\partial\zeta}(\xi_j,\zeta_k) &= \sum_{m=0}^{M}\sum_{n=0}^{N} \phi_{mn} g_m(\xi_j) h'_n(\zeta_k),\end{aligned} \tag{30}$$

and similarly for higher derivatives. See [11] for example, for a detailed description of this approach.

The known nodal values of the velocities on the boundaries $j = 0$ and $j = M$ are incorporated explicitly in (29) and (30). This leaves $[7(N+1) - 2] \times (M+1)$ unknown nodal values to be determined at each time step. This is achieved by pointwise collocation of all variables and their spatial derivatives in equations (2)-(7), using the following time stepping scheme:

$$\frac{\mathbf{T}^{n+1} - \mathbf{T}^n}{\Delta t} = \frac{2\eta^n}{\lambda^n}\mathbf{d}^n - \frac{\mathbf{T}^n}{\lambda^n} - \mathbf{v}^n \cdot \nabla\mathbf{T}^n + (\nabla\mathbf{v}^n)\cdot\mathbf{T}^n + \mathbf{T}^n\cdot(\nabla\mathbf{v}^n)^T; \tag{31}$$

$$\frac{\mathbf{v}^* - \mathbf{v}^n}{\Delta t} = \frac{1}{\rho}\nabla\cdot\mathbf{T}^n - \mathbf{v}^n\cdot\nabla\mathbf{v}^n; \tag{32}$$

$$\frac{\mathbf{v}^{n+1} - \mathbf{v}^*}{\Delta t} = -\frac{1}{\rho}\nabla p^{n+1}, \qquad \nabla\cdot\mathbf{v}^{n+1} = 0; \tag{33}$$

$$\frac{\Theta^{n+1} - \Theta_n}{\Delta t} = -\mathbf{v}^n\cdot\nabla\Theta^n + \frac{\kappa}{\rho C_p}\nabla^2\Theta^n + \frac{1}{\rho C_p}\Phi^n, \tag{34}$$

where Δt denotes the time step.

Equations (31) and (34) are explicit formulae for evaluating $\mathbf{T}$ and Θ at time-level $n+1$ from knowledge of all the variables and their derivatives at time-level n. Equations (32) and (33) split the solution of the momentum equations into two stages: an explicit stage which defines an intermediate non-divergence-free velocity $\mathbf{v}^*$ and an implicit stage which determines simultaneously a pressure field and a corrected divergence-free velocity at time-level $n+1$.

Care has to be taken in solving (33) since the underlying pressure matrix is singular. One zero singular value arises because the pressure is underdetermined (up to a constant), but another zero singular value arises due to the fact that the pressure and velocities are determined at the same grid points, thereby including a spurious pressure mode. This situation is easily remedied using singular value decomposition. The details may be found in [10]. The singular value decomposition need only be carried out once since the pressure matrix in (33) is independent of time.

The accuracy is spectral in space but only first-order in time. More detailed studies of the time-stepping scheme and a modification may be found in [15]. Very small time steps (down to 10^{-9}) are needed for convergence at high eccentricities and/or high relaxation times. This disadvantage is offset by ease of programming, with all stages apart from (33) being explicit.

5. Numerical results

TABLE I
Geometrical data and fluid parameters

R_J	=	3.125×10^{-2} (m)
c	=	4.000×10^{-5} (m)
Ω	=	$500(s^{-1}) \approx 4775(rpm)$
ρ	=	8.200×10^{2} (kgm^{-3})
C_p	=	1.750×10^{3} $(Jkg^{-1}K^{-1})$
κ	=	1.4×10^{-1} $(Wm^{-1}K^{-1})$
η_∞	=	4.500×10^{-4} (Pas)
η_0	=	9.352×10^{-4} (Pas)
$\bar{\alpha}$	=	2.390×10^{-8} (Pa^{-1})
$\bar{\beta}$	=	2.575×10^{3} (^0K)
α	=	1.119×10^{-8} (Pa^{-1})
β	=	7.580×10^{2} (^0K)
m	=	0.5450
E	=	-21.33
k_t	=	2.50×10^{-4}

The geometric data and fluid parameters used in our numerical experiments are listed in Table I. Wherever possible the values in Table I are taken from existing physical configurations, or are best-fitted to experimental data in [2, 7]. The constant k_t in equation (14) is chosen to give a relaxation time of 2×10^{-6}s for the zero shear-rate viscosity at a temperature of 80^0C and an effective pressure of 1MPa. We note

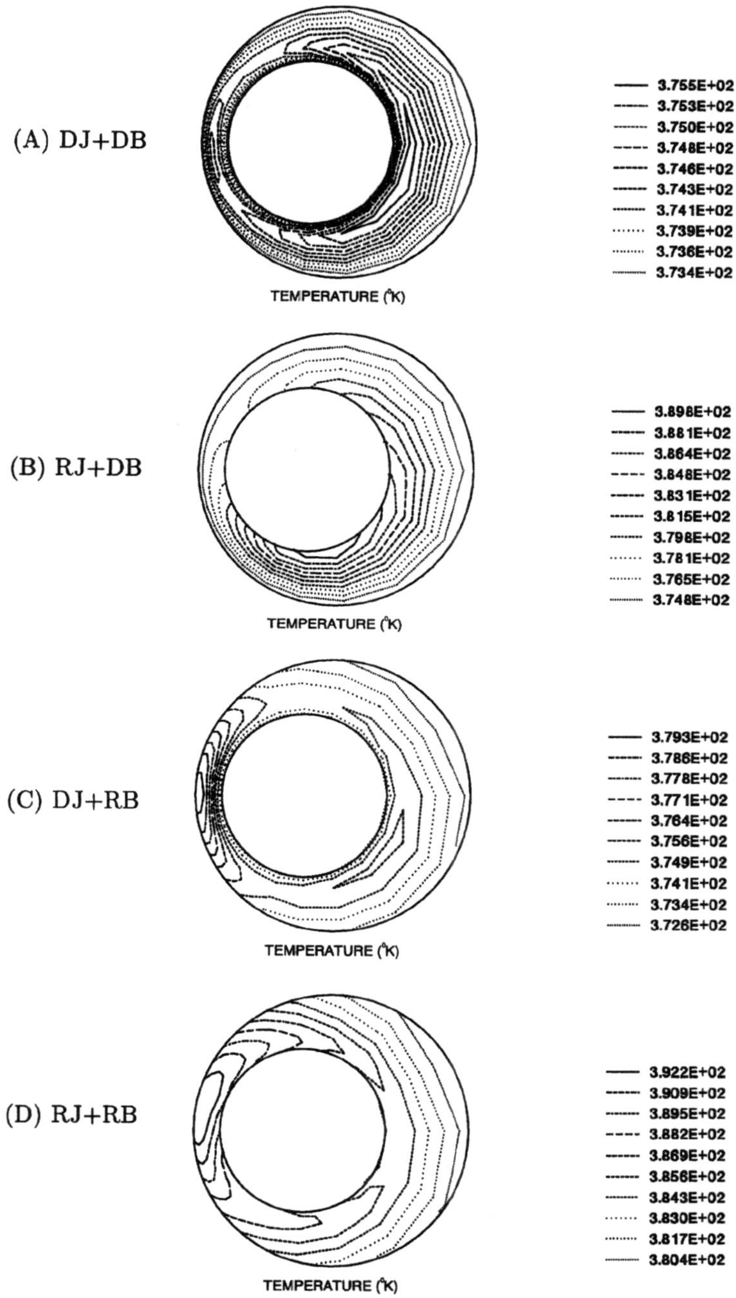

Fig. 2. Contours of the temperature field calculated by using different thermal boundary conditions. Plots are of (A) DJ+DB, (B) RJ+DB, (C) DJ+RB, (D) RJ+RB. All calculations are for $\epsilon = 0.75$ and $\Theta_J = \Theta_B = 373.15$, $\Theta_{rJ} = \Theta_{rB} = 368.15^0 K$ and $Bi = 1.0$.

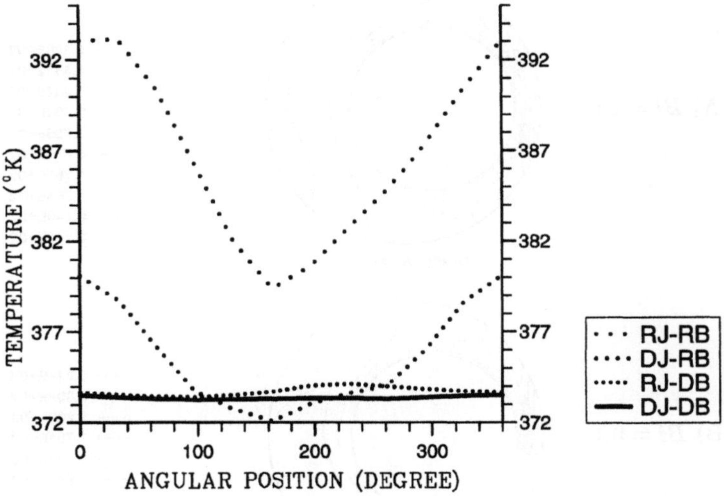

Fig. 3. Comparison of the temperature close to the outer cylinder in the azimuthal direction with the different thermal boundary conditions at $\epsilon = 0.75$. All calculations are for $\Theta_J = \Theta_B = 373.15$, $\Theta_{rJ} = \Theta_{rB} = 368.15^0 K$ and $Bi = 1.0$.

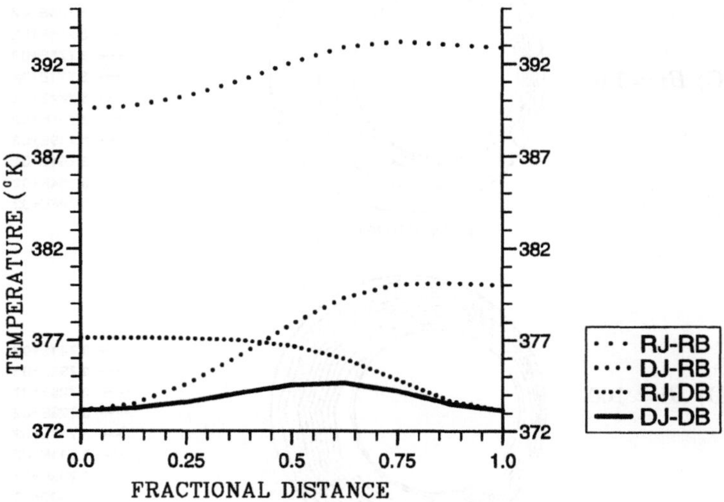

Fig. 4. Comparison of the temperature across the smallest gap from the inner to the outer cylinders with the different thermal boundary conditions at $\epsilon = 0.75$. All calculations are for $\Theta_J = \Theta_B = 373.15$, $\Theta_{rJ} = \Theta_{rB} = 368.15^0 K$ and $Bi = 1.0$.

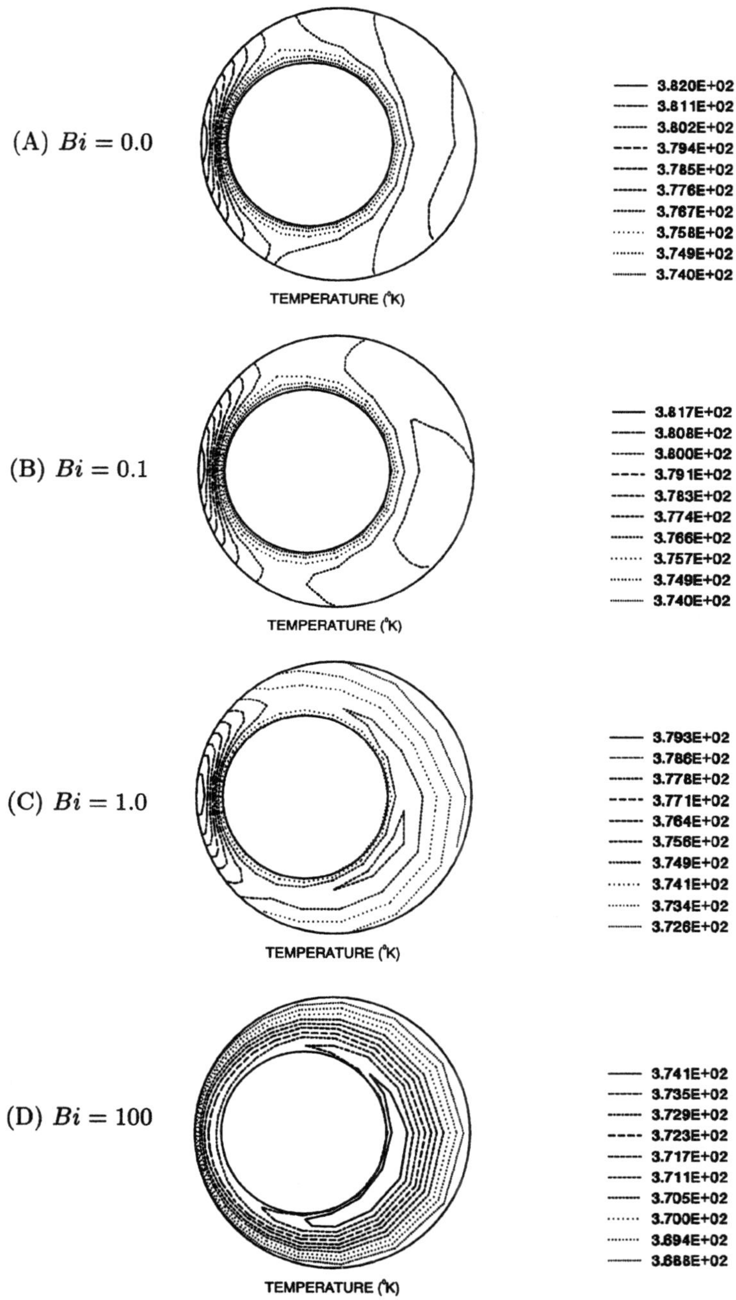

Fig. 5. Contours of the temperature field calculated by using the thermal boundary condition DJ+RB with the different Biot numbers. All calculations are for the eccentricity $\epsilon = 0.75$ and $\Theta_J = 373.15^0 K, \Theta_{rB} = 368.15^0 K$.

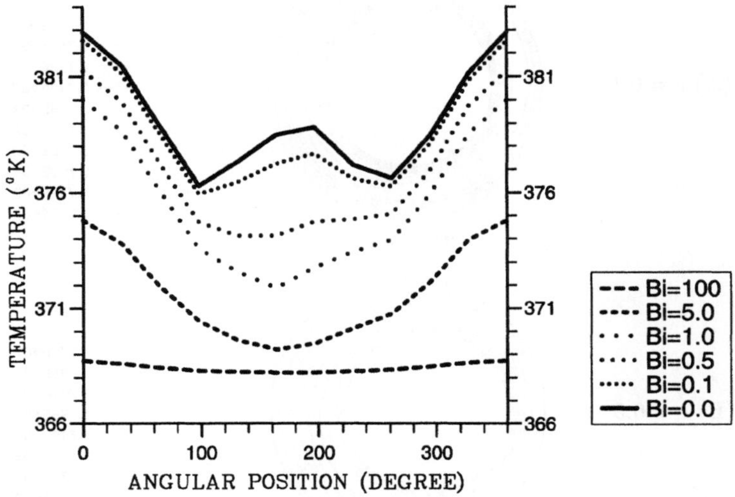

Fig. 6. Effect of increasing the Biot numbers on the profile of the temperatures close to the outer cylinder in the azimuthal direction for $\epsilon = 0.75$ and $\Theta_J = 373.15^0 K, \Theta_{rB} = 368.15^0 K$.

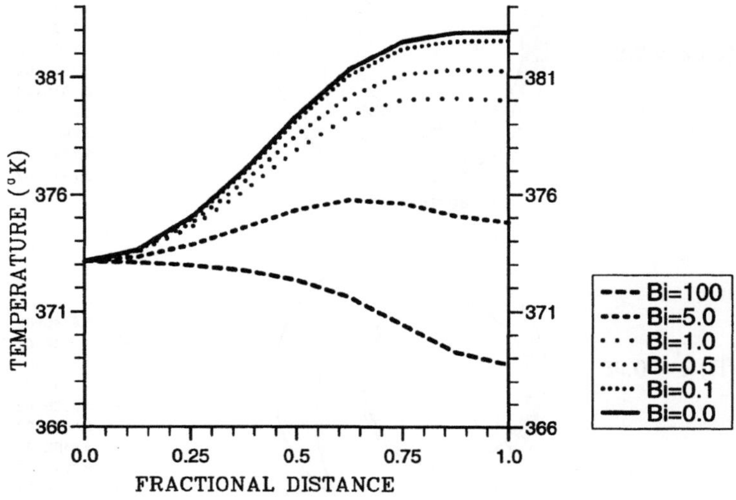

Fig. 7. Effect of increasing the Biot numbers on the profile of the temperatures across the smallest gap from the inner to the outer cylinders for $\epsilon = 0.75$ and $\Theta_J = 373.15^0 K, \Theta_{rB} = 368.15^0 K$.

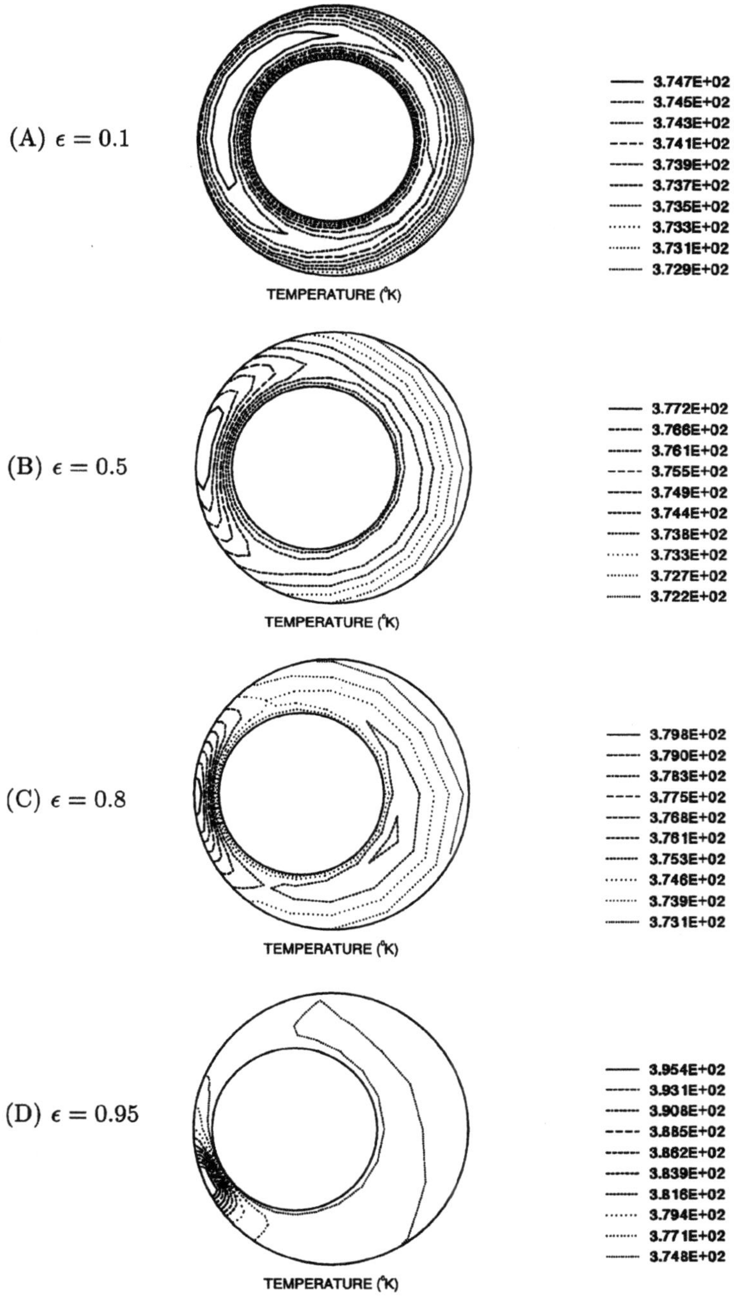

Fig. 8. Contours of the temperature field calculated by using the thermal boundary condition DJ+RB with the different eccentricities ϵ. All calculations are for $\Theta_J = 373.15^0 K$, $\Theta_{rB} = 368.15^0 K$ and $Bi = 1.0$.

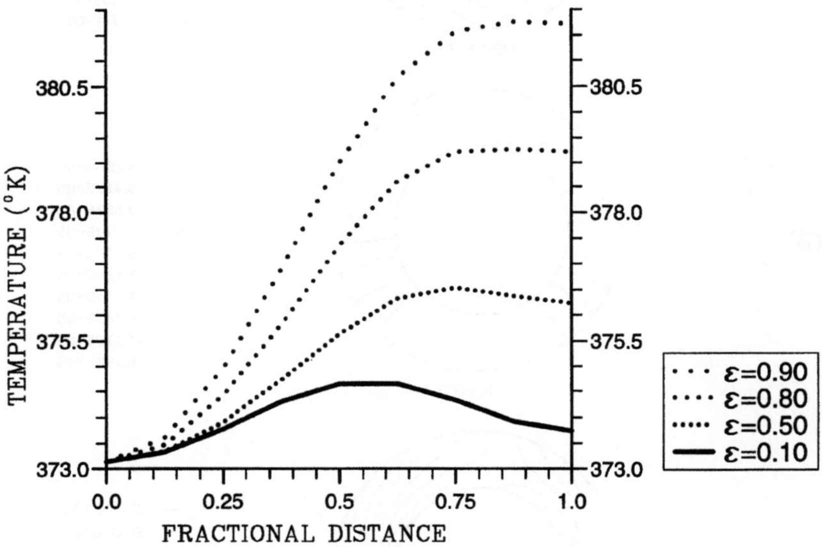

Fig. 9. Effect of increasing the eccentricity ϵ on profile of the temperatures across the smallest gap from the inner to the outer cylinders with the thermal boundary condition DJ+RB for $\Theta_J = 373.15^0 K$ and $\Theta_{rB} = 368.15^0 K$.

(A)

(B)

(C)

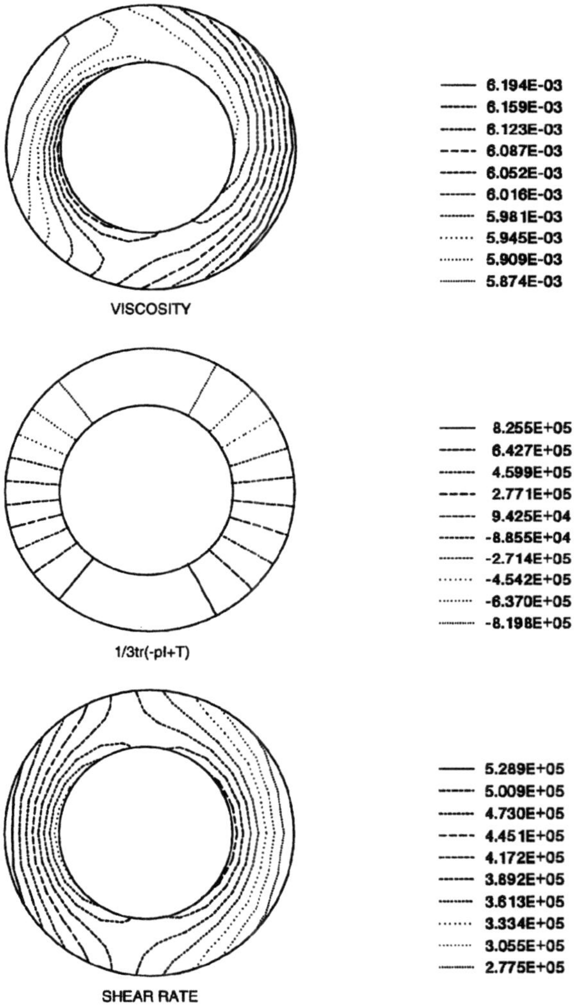

Fig. 10. Contours of viscosity, shear rate pressure and the trace of the total stress with $\epsilon = 0.1$. Plots are of (A) viscosity filed, (B) $\frac{1}{3}$trace $(-p\mathbf{I} + \mathrm{T})$, (C) shear rate.

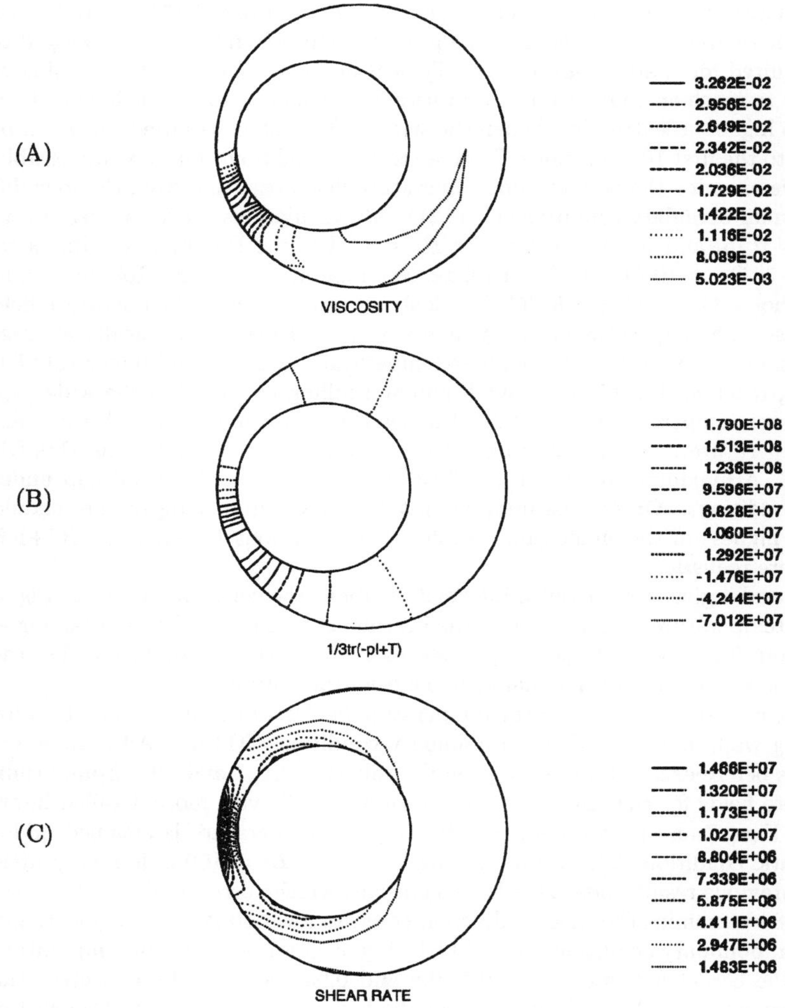

Fig. 11. Contours of viscosity, shear rate pressure and the trace of total stress with $\epsilon = 0.95$. Plots are of (A) viscosity filed, (B) $\frac{1}{3}$trace $(-p\mathbf{I} + \mathbf{T})$, (C) shear rate.

that in this paper we have not attempted to model extensional effects, and have therefore taken $K_e = 0$.

In each experiment the time stepping algorithm is run from a state of rest until an equilibrium state is reached. Time steps Δt in the range 10^{-6}s to 10^{-9}s are required for temporal convergence, while grid numbers M and N in the range 8 to 16 are required for spatial convergence. Typical physical time (i.e. $\Delta t \times$number of iterations) to reach an equilibrium state varies from about 1s to 6s, the longer times relating to higher eccentricities. When the smallest time steps are used, they can be restricted to the first 10^6 iterations. Thereafter, increased time steps are admissable.

First we examine the equilibrium temperature fields resulting from the four different thermal boundary conditions (18)-(21). All calculations are for a fixed eccentricity $\epsilon = 0.75$, and fixed Biot number $Bi = 1$. For the Dirichlet conditions we choose $\Theta_J = \Theta_B = 100^0C$, while for reference temperatures for the Robbins conditions we choose $\Theta_{rJ} = \Theta_{rB} = 95^0C$. Fig. 2 shows contours of the temperature field in each case. The gap between the cylinders is exaggerated schematically for ease of illustration. We note that the highest temperatures are attained under RJ+RB conditions, reaching 119^0C in the small gap and falling to 107^0C in the wide gap. Slightly lower temperatures are attained under RJ+DB conditions, with the maximum temperature occuring away from the small gap. Similarly for the DJ+DB conditions. A maximum temperature of 106^0C is reached in the small gap under DJ+RB conditions. On the assumption that the equilibrium temperature should always be highest in the small gap, we infer that conditions DJ+RB and RJ+RB are the more realistic.

Figs. 3 and 4 give more detailed information about the contours in Fig. 2. Fig. 3 shows the temperature profiles close to the bearing as a function of azimuthal angle, starting from 0 at the small gap. Fig. 4 shows the temperature profiles across the small gap as a function of fractional distance from the journal.

Next we consider the effect on the temperature field of different Biot numbers for the bearing wall, under the thermal boundary conditions DJ+RB. All parameters are kept as before except that Bi is given the values 0, 0.1, 1 and 100. Equilibrium temperature fields for the four cases are shown in Fig. 5, with more detailed information in Figs. 7 and 6. We note that the highest temperature is attained under the adiabatic condition (17) on the bearing. The case $Bi = 100$ differs very little from the previous result under DJ+DB conditions, which corresponds to $Bi \to \infty$.

In Fig. 8 we examine the effect of different eccentricities. Bi is kept fixed at 1, and the thermal boundary conditions are DJ+RB. Fig. 8 shows equilibrium temperature fields for the eccentricities $\epsilon = 0.1$, 0.5, 0.8 and 0.95. As is to be expected, the highest temperature (122^0C) is attained for the highest value of ϵ. Fig. 9 shows the temperature profile across the small gap for different eccentricities.

Finally, we examine the effect on other fields of the thermal boundary condition DJ+RB. In Figs. 10 and 11 we show contours of viscosity, effective pressure and shear rate, for small and large eccentricities. In Fig. 10, where $\epsilon = 0.1$, we note very little variation in viscosity and shear rate, whereas in Fig. 11, where $\epsilon = 0.95$, the viscosity and shear rate change by an order of magnitude. Correspondingly, we obtain a change in relaxation time of an order of magnitude, which gives a possible indication that the resulting increase in normal stress may prove beneficial with regard to load-bearing capacity. This issue is therefore worth further study.

It is also worth noting the difference in effective pressures in Figs. 10 and 11. When $\epsilon = 0.1$ the maximum pressure is just under 1MPa and the pressure field is almost perfectly antisymmetric. For $\epsilon = 0.95$, the maximum pressure increases to almost 180MPa, and the field loses its antisymmetry. This trend away from negative pressures means that the fluid is less likely to cavitate. It is again worth further study to establish whether this is a positive benefit of viscoelasticity.

Acknowledgements

This work was funded by Shell Research Ltd., Thornton, UK. We have benefited from collaboration with Drs. T. Bates, R. Coy, L. Scales and A. Schlijper at Shell Thornton, and with Drs. D.R. Gwynllyw, T.N. Phillips, G.W. Roberts and Professor K. Walters at the University of Wales. The nonisothermal code we have used in this work was adapted from an earlier isothermal code written by G.W.Roberts, to whom we are greatly indebted.

References

1. H. A. Barnes, J. F. Hutton and K. Walters, *An introduction to rheology*. Elsevier, 1989.
2. T.W. Bates, B. Williamson, J.A. Spearot and C.K. Murphy, A correlation between engine oil rheology and oil film thickness in engine journal bearing. *Society of Automotive Engineers*, 1986, Paper No. 860376.
3. A.N. Beris, R.C. Armstrong and R.A. Brown, Spectral/finite element calculations of the flow of a Maxwell fluid between eccentric rotating cylinders. *J. Non-Newtonian Fluid Mech.* 22(1987)129-167.
4. A.N. Beris, R.C. Armstrong and R.A. Brown, Finite element calculation of viscoelastic flow in a journal bearing. Part I. *J. Non-Newtonian Fluid Mech.* 16(1984)141-172. Part II. *J. Non-Newtonian Fluid Mech.* 19(1986)323-347.
5. A.J. Chorin, Numerical solution of the Navier-Stokes equations. *Math. Comput.*, 22 (1968) 745-762.
6. T. Han and R. S. Paranjpe, A finite volume analysis of the thermohydrodynamic performance of finite journal bearings. *ASME Journal of Tribology*, 112 (1990) 557-566.
7. J.F. Hutton,B. Jones and T.W. Bates, Effects of isotropic pressure on the high temperature high shear viscosity of motor oils. *Society of Automotive Engineers*, 1983, Paper No. 830030.
8. H. H. Ott and G. Paradissiadis, Thermohydrodynamic analysis of journal bearings considering cavitation and reverse flow. *ASME Journal of Tribology*, 110 (1988) 439-447.
9. G.W. Roberts and K. Walters, On viscoelastic effects in journal-bearing lubrication. *Rheologica Acta*, 31 (1992) 55-62.
10. T.N. Phillips and G.W. Roberts, The treatment of spurious pressure modes in spectral incompressible flow calculations. *J. Comput. Phys.*, 105 (1993) 150-165.
11. R.G. Voigt, D. Gottlieb and M.Y. Hussaini, *Spectral methods for partial differential equations*. SIAM, Philadelphia, 1984.
12. K. Walters, Polymers as additives in lubricating oils. *Polyméres et lubrification*, CNRS Paris, (1974) 27-36.
13. J.L. White and A.B. Metzner, Development of constitutive equations for polymeric melts and solutions, *J. Appl. Polym. Sci.* 7 (1963)1867-1889.
14. H.H. Winter, Viscous dissipation in shear flows of molten polymers, *Adv. Heat Transfer*, 13 (1977) 205-267.
15. S. Xu, A.R. Davies and T.N. Phillips, Pseudospectral method for transient viscoelastic flow in an axisymmetric channel. *Numerical Methods for Partial Differential Equations*, 9 (1993) 691-710.

NUMERICAL ANALYSIS OF THE NON-ISOTHERMAL VISCOELASTIC FLOW OF A POLYMER MELT AROUND A CYLINDER

FRANK P.T. BAAIJENS
Philips Research Laboratories,
P.O. Box 80000, 5600 JA Eindhoven, The Netherlands
and
Eindhoven University of Technology
Center for Polymers and Composites
P.O. Box 513, 5600 MB Eindhoven, The Netherlands

Abstract.
The non-isothermal steady viscoelastic flow of a polystyrene melt around a cylinder is analysed numerically. In particular, the effect of viscous dissipation on the stress distribution in the vicinity of the cylinder is investigated and compared with isothermal predictions.

A low-order discontinuous Galerkin method is used for the advective parts of the constitutive equation, while the streamline upwind Petrov Galerkin method is used to discretize the energy equation. A fully coupled Newton iteration procedure is applied to solve the resulting non-linear set of equations.

The material is modelled with a Phan-Thien Tanner model.

Key words: Viscoelastic flow, non-isothermal, Discontinuous Galerkin method, stabilization.

1. Introduction

In Baaijens et al. [1], the flow of a polyisobutylene solution in tetradecane (PIB/C14) is investigated numerically and experimentally. A striking difference is found between the predicted and measured (with flow-induced birefringence) stress profiles when using the Phan-Thien Tanner (PTT) equation to model the fluid rheology. It is interesting to check the performance of the PTT model for polymer melts as well. One of the polymers to be investigated experimentally in the near future, is a commercial polystyrene (PS674E from DOW Chemical), whose viscometric and thermal characterization may be found in Douven [2]. Given the high viscosity of these materials under normal processing conditions, viscous dissipation in the experimental setup may be expected. The purpose of this work is to numerically investigate the effects of this viscous dissipation on the stress distribution near the cylinder.

Clearly, many of the intrinsic difficulties of isothermal viscoelastic flows are also present under non-isothermal flow conditions. In this work a mixed stress-velocity-pressure-temperature formulation is used. A particularly difficult task is to define well-balanced discrete approximations of the stress and velocity spaces. As argued in Baaijens [3], the compatibility requirements on the stress-velocity discretization are most easily met by using discontinuous approximations of the extra stress tensor. The advective parts of the constitutive equation then necessitate the use of the discontinuous Galerkin method (DG). This scheme, however, may produce os-

J.F. Dijksman and G.D.C. Kuiken (eds.),
IUTAM Symposium on Numerical Simulation of Non-isothermal Flow of Viscoelastic Liquids, 179-190.

cillatory stress fields near singularities or strong stress boundary layers if high-order approximations are used, see Fortin and Fortin [4], Fortin et al. [5] and Baaijens [3]. This difficulty can be overcome by using a piecewise constant approximation of the extra stress tensor, see Baaijens [3]. However, this necessitates the use of a special procedure to discretize the velocity and pressure spaces. The work of Franca and Stenberg [6] is followed closely on this matter.

The Streamline Upwind Petrov Galerkin (SUPG) method, originally proposed by Brooks and Hughes [7], is used to discretize the energy equation.

A fully coupled Newton iteration procedure is applied to solve the resulting set of non-linear equations, while the Jacobian matrix is found by one-sided finite differencing to provide for maximum flexibility during the investigation of various constitutive models.

This paper is constructed as follows. First the problem is defined, following which the numerical method is explained in detail. The geometric and boundary condition assumptions are discussed next. Then the viscometric and thermal characteristics of the polystyrene melt are reviewed. A one-mode Phan-Thien Tanner model is used to fit the viscometric data, and time-temperature superposition is assumed to hold. Computational results for the non-isothermal analysis are compared with the isothermal case, showing significant viscous heating at moderated values of the Deborah number having a noticeable impact on the stress distribution. Finally, conclusions are drawn.

2. Problem definition

Under the experimentally possible processing conditions, the fluid may be assumed to be inertialess and incompressible, and the gravitational effects are neglected. Consequently, the momentum equation and continuity equation reduce to

$$\vec{\nabla} \cdot (-p\boldsymbol{I} + \boldsymbol{\tau}) = \vec{0} \tag{1}$$

$$\vec{\nabla} \cdot \vec{u} = 0 \tag{2}$$

where p is the pressure, $\boldsymbol{\tau}$ the extra stress tensor and $\vec{u}$ the velocity field.

To model the fluid rheology, the Phan-Thien Tanner (PTT) model is used

$$\overset{\nabla}{\boldsymbol{\tau}} + (\frac{1}{\lambda} + \frac{\varepsilon}{\eta}\mathrm{tr}(\boldsymbol{\tau}))\boldsymbol{\tau} = \frac{2\eta}{\lambda}\boldsymbol{D} \tag{3}$$

In this, λ represents the relaxation time, η the viscosity and ε is a dimensionless parameter that affects the shear thinning and, predominantly, the elongational properties of the PTT model. The strain rate tensor is defined by $2\boldsymbol{D} = \boldsymbol{L} + \boldsymbol{L}^T$ with $\boldsymbol{L} = (\vec{\nabla}\vec{u})^T$ the velocity gradient tensor. Under steady flow conditions

$$\overset{\nabla}{\boldsymbol{\tau}} = \vec{u} \cdot \vec{\nabla}\boldsymbol{\tau} - \boldsymbol{L} \cdot \boldsymbol{\tau} - \boldsymbol{\tau} \cdot \boldsymbol{L}^T \tag{4}$$

The viscosity η and relaxation time λ are temperature dependent. Time-temperature superposition is assumed to hold, such that

$$\eta(T) = a_T\eta(T_0) \ , \quad \theta(T) = a_T\theta(T_0) \tag{5}$$

with T the temperature, T_0 a reference temperature and a_T the shift-factor that obeys the WLF equation

$$log(a_T) = \frac{-c_1(T - T_0)}{c_2 + T - T_0} \tag{6}$$

The energy equation is approximated by

$$\rho c_p \vec{u} \cdot \vec{\nabla} T = \vec{\nabla} \cdot (\kappa \vec{\nabla} T) + \boldsymbol{\tau} : \boldsymbol{D} \tag{7}$$

where ρ is the density, c_p the specific heat at constant pressure and κ the thermal conductivity coefficient. Eq. 7 is only an approximation to the energy equation because, among other reasons, $\boldsymbol{\tau} : \boldsymbol{D}$ does not exactly represent the viscous dissipation, see Peters [8]. The extra stress tensor $\boldsymbol{\tau}$ consists of both a reversible and an irreversible part. The irreversible part should, formally, not be taken into account. However, it is commonly believed ([9]) that eq. 7 is a reasonable approximation.

As boundary conditions, a fully developed flow is assumed at the inflow boundary. There, velocities, extra stresses and temperature are prescribed. At other parts of the boundary, velocities, stresses, temperatures and temperature gradients may be prescribed.

3. Numerical method

There are a number of difficulties associated with the finite element solution to the previous problem when cast in a mixed formulation.

First of all, the discretization spaces of the primary variables, pressure, velocity and stresses, need to be matched carefully. It is well known, for two-dimensional problems at least, how the discrete pressure and velocity space are to be matched. Numerous elements exist that satisfy the so-called inf-sup compatibility condition on the velocity-pressure approximation, see Pironneau [10].

Much less is known about the selection of the discrete stress space with respect to the velocity space. From the work of Franca and Stenberg [6] it follows that the inf-sup compatibility requirement of the velocity-stress interpolation is easiliy satisfied if discontinuous stresses are used. For continuously interpolated stresses the situation is more complicated. For instance, to obtain stable stress fields for the Stokes problem, Marchal and Crochet [11] used a four-by-four bilinear stress subelement division on each quadrilateral velocity element. However, Baranger and Sandri [12] conclude that an equal order interpolation of the stresses and velocity is formally admissible at non-zero solvent viscosity. This is not supported by the computational results of Marchal and Crochet [11].

Secondly, both the constitutive model and the energy equation contain convective terms. It is common practice to apply a continuous interpolation of the extra stresses and use Streamline Upwind Petrov Galerkin (SUPG), [13], [14], [15], [16], [17], [18] and [19], or just (inconsistent) Streamline Upwinding (SU), Marchal and Crochet [11]. For problems containing singularities, Marchal and Crochet [11] demonstrated that SUPG gives insufficient stability compared with the SU method. The latter, though, may give rise to significant consistency errors, see Tanner and Jin [20]. The Discontinuous Galerkin (DG) method as proposed for viscoelastic flows by Fortin and Fortin [4], also produces oscillatory results near singularities or strong stress boundary layers, see Baaijens [21].

Most likely, the oscillations found with the SUPG and DG method are due the fact that these methods are not Total Variation Diminishing (TVD), see van Leer [22]. It is for this reason that Baaijens [19] investigated discontinuity capturing to stabilize the stress field near the corner singularity in the four-to-one contraction problem, although with limited success. The use of monotonicity enforcement as proposed by van Leer [22], in combination with the DG method on piecewise linear discontinuous stress fields, leads to better results, see Baaijens [21], but only applies to unsteady problems.

A more fruitful path appears to be the use of a piecewise-constant discontinuous extra-stress interpolation in combination with the DG method. However, in this case, in order to have compatibility between the stress and velocity discretization, a linear interpolation of the velocity space on triangles is mandatory. Notice that in this case the requirement set by Marchal and Crochet [11], namely that the stress space must be a subset of the velocity gradient space, is identically satisfied. Furthermore, a linear continuous pressure discretization is employed that is stabilized by using the Galerkin Least Squares methodology recently developed by Franca and Stenberg [6]. They introduced a stabilized stress-velocity-pressure formulation for the Stokes problem that allows various combinations of interpolation schemes of these variables, see also Hughes et al. [23], Franca and Hughes [24], Tezduyar et al. [25] and Behr et al. [26]. In particular, it allows an equal-order interpolation of the velocity and pressure field. Using a linear interpolation on triangles for these fields, it is possible to use a piecewise-constant discontinuous interpolation of the extra stress tensor.

The temperature field is approximated with a continuous linear field on triangles. Define $\mathcal{L}\boldsymbol{\tau}$ as

$$\mathcal{L}\boldsymbol{\tau} = \vec{u} \cdot \vec{\nabla}\boldsymbol{\tau} - \boldsymbol{L} \cdot \boldsymbol{\tau} - \boldsymbol{\tau} \cdot \boldsymbol{L}^T + (\frac{1}{\lambda} + \frac{\varepsilon}{\eta}\mathrm{tr}(\boldsymbol{\tau}))\boldsymbol{\tau} \tag{8}$$

and set

$$\vec{\mathcal{R}} = \vec{\nabla} \cdot (-p\boldsymbol{I} + \boldsymbol{\tau}) \tag{9}$$

Then, assuming a discontinuous interpolation of the extra stress tensor from the onset, the application of the Discontinuous Galerkin (DG) method and the generalization of the stability concepts of Franca and Stenberg [6], and the use of SUPG for the energy equation, give

Problem 1 (DG) *Find $(\vec{u}, p, \boldsymbol{\tau}, T)$, such that for all admissible $(\vec{v}, q, \boldsymbol{s}, r)$*

$$-(\boldsymbol{D}_v, \boldsymbol{\tau}) + (\vec{\nabla} \cdot \vec{v}, p) = 0 \tag{10}$$

$$(q, \vec{\nabla} \cdot \vec{u}) - \sum_{e=1}^{N_{el}} (\delta \vec{\nabla} q, \vec{\mathcal{R}})_e = 0 \tag{11}$$

$$(\boldsymbol{s}, \mathcal{L}\boldsymbol{\tau} - 2\frac{\eta}{\lambda}\boldsymbol{D}_u) + \sum_{e=1}^{N_{el}} \left[(\delta \vec{\nabla} \cdot \boldsymbol{s}, \vec{\mathcal{R}})_e - \int_{\Gamma_{in}^e} \boldsymbol{s} : \vec{u} \cdot \vec{n}(\boldsymbol{\tau} - \boldsymbol{\tau}^{ext})\, d\Gamma \right] = 0 \tag{12}$$

$$(\tilde{r}, \rho c_p \vec{u} \cdot \vec{\nabla} T - \boldsymbol{\tau} : \boldsymbol{D}_u) + (\vec{\nabla} \tilde{r}, \kappa \vec{\nabla} T) = 0 \tag{13}$$

with

$$\tilde{r} = r + \alpha \vec{u} \cdot \vec{\nabla} r \ , \quad \alpha = \left[\frac{\partial \xi_i}{\partial x_j} \frac{\partial \xi_i}{\partial x_k} u_j u_k \right]^{-1/2} \ , \quad i = 1,2,3 \quad j,k = 1,2 \tag{14}$$

with ξ_i *being the* $i-th$ *local 'triangle' coordinates of the element.*

In this, $(.,.)$ and $(.,.)_e$ denote the appropriate inner product on Ω and Ω^e, respectively; Γ^e_{in} is the inflow boundary of element Ω^e, $\vec{n}$ the unit outward normal to Γ^e, $\boldsymbol{\tau}^{ext}$ the stress tensor of the neighbouring upwind element, δ is a stability constant as defined below, and $\boldsymbol{D}_u = \boldsymbol{D}(\vec{u})$, and $\boldsymbol{D}_v = \boldsymbol{D}(\vec{v})$.

The parameter δ is defined by

$$\delta = \frac{h_e^2}{32\bar{\eta}} \ , \quad \bar{\eta} = \frac{\eta^2}{\eta + \varepsilon \lambda \text{tr}(\boldsymbol{\tau})} \ , \qquad h_e = \sqrt{area(\Omega^e)}. \tag{15}$$

Due to the piecewise constant approximation of the extra stress tensor $\boldsymbol{\tau}$, and its associated weight function $\boldsymbol{s}$,

$$\vec{\nabla} \cdot \vec{\mathcal{R}} = \vec{0} \tag{16}$$

while due to the linear interpolation of the velocity field

$$\vec{\nabla} \cdot \boldsymbol{D}_v = \vec{0} \tag{17}$$

Furthermore, the only remaining term in $\vec{\mathcal{R}}$ is

$$\vec{\mathcal{R}} = \vec{\nabla} \cdot (-p\boldsymbol{I}) \tag{18}$$

Consequently, the terms involving δ in eq. 10 and 12 vanish, while eq. 11 reduces to

$$(q, \vec{\nabla} \cdot \vec{u}) + \sum_{e=1}^{N_{el}} (\delta \vec{\nabla} q, -\vec{\nabla} p)_e = 0 \tag{19}$$

This is a simple modification of the regular formulation.

After discretization, a nonlinear set of equations results, which is solved with a fully coupled Newton iteration scheme. The Jacobian matrix is obtained by one-sided finite differencing. Compared to an analytically obtained Jacobian, this provides for enhanced flexibility in experimenting with different constitutive models, at the expence of increased computational cost.

4. Model definition

Geometry definition. The flow around a cylinder placed between two parallel plates is investigated, see Fig. 1. The cylinder has radius $R = 4\ 10^{-3}$ m and the half channel height is denoted by $H = 8\ 10^{-3}$ m. The computational domain covers $-10 < x/R < 10$

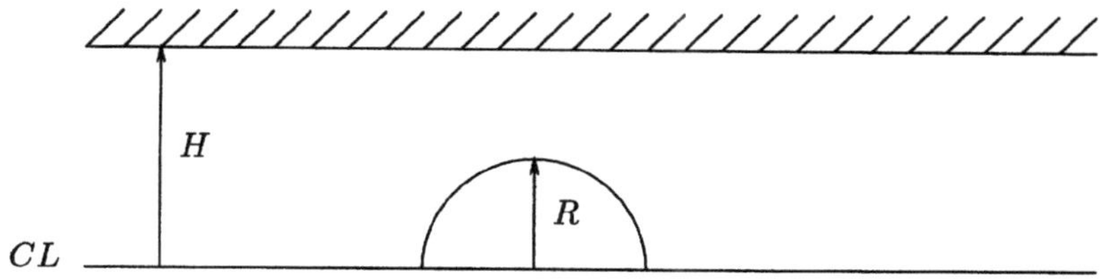

Fig. 1. Geometry of flow around a cylinder placed between two parallel plates

λ (s)	0.15
η (Pas)	10^4
ε	0.1
ρ (kg /m^3)	921
c_p (kJ/(kg K)	1.5
κ (W/(mK)	0.17
c_1	4.54
c_2 (K)	150.36
T_0 (K)	462

TABLE I
Material parameters for PS 678E Dow Chemical

Boundary conditions. Along the entrance steady, fully developed isothermal flow conditions hold. The wall and entrance temperatures are 462 K. The channel walls can be thermostated relatively well. The glass windows on either side of the channel to allow a laser to pierce through the melt, isolating the cavity. Given the small radius of the cylinder, it is unfeasible to separately cool, or heat, the cylinder. As the width of the channel is at least ten times the height of the channel, it assumed that the cylinder is isolated, hence a vanishing normal temperature gradient $(\vec{n}\cdot\vec{\nabla}T)$ is assumed at the cylinder wall.

Material properties. The material properties are found in Douven [2]. A one-mode fit to the viscometric data is made, giving the relaxation time and viscosity of table I. Thermal conductivity, density and specific are temperature dependent, but in the range of temperatures of interest here, they may be approximated by constants, see table I. Furthermore, the WLF parameters are given.

5. Results

To facilitate the discussion of the results, three dimensionless groups are introduced: the Deborah (De), Peclet (Pe) and the heat generation number (Gn). By scaling the velocity with the average upstream velocity V, length scales with the radius

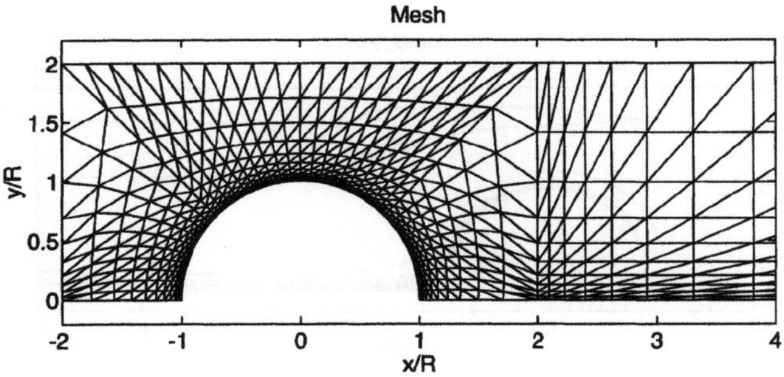

Fig. 2. Detail of the mesh near the cylinder.

of the cylinder R, stresses and pressure with $\eta_0 V/R$ and the temperature with the reference temperature T_0, the PTT model and the energy equation may be rewritten as

$$De\, \overset{\nabla}{\boldsymbol{\tau}^*} + (1 + \varepsilon De\, \mathrm{tr}(\boldsymbol{\tau}^*))\boldsymbol{\tau}^* = \boldsymbol{D}^* \tag{20}$$

$$Pe\, \vec{u}^* \cdot \vec{\nabla}^* T^* = \vec{\nabla}^* \cdot \vec{\nabla}^* T^* + Gn\, \boldsymbol{\tau}^* : \boldsymbol{D}^* \tag{21}$$

with

$$\boldsymbol{\tau} = \frac{\eta_0 V}{R}\boldsymbol{\tau}\,, \quad \vec{u} = V\vec{u}^*\,, \quad \boldsymbol{D} = \frac{V}{R}\boldsymbol{D}^*\,, \quad T = T_0\, T^* \tag{22}$$

and

$$De = \frac{\lambda_0 V}{R}\,, \quad Pe = \frac{\rho c_p V R}{\kappa}\,, \quad Gn = \frac{\eta_0 V^2}{\kappa T_0} \tag{23}$$

The viscosity η_0 and the relaxation time λ_0 are the viscosity η and relaxation time λ at the reference temperature T_0, respectively, see table I.

The maximum average velocity used in the computations is $V = 0.07m/s$. Hence $De_{max} = 2.975$, $Pe_{max} = 2100$ and $Gn_{max} = 0.53$. Clearly, the thermal problem is convection dominated, with some heat generation.

Part of the mesh used in the computations is shown in Fig. 2.

At $De = De_{max}$, Fig. 3 shows the temperature rise with respect to the reference temperature $(T - T_0)$ in the vicinity of the cylinder. A strong temperature boundary layer is formed at the downstream section due to viscous heating. Not only the shear flow at the edge of the cylinder, but also the elongational flow at the tail of the cylinder causes significant heating. At this Deborah number a temperature rise of

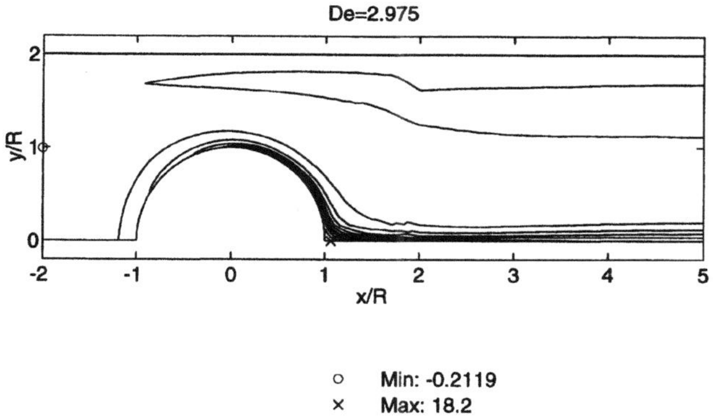

Fig. 3. Distribution of the temperature rise $T - T_0$. At this Deborah number a temperature rise of more than 18 K is seen.

about 18 K is observed. This is also illustrated by Fig. 4, showing the temperature rise along the symmetry line at four values of the Deborah number. The amount of temperature rise is almost a linear function of the Deborah number.

A temperature rise of 18 K has a significant effect on the stress levels near the cylinder. This is shown in Fig. 5 where the dimensionless τ^*_{xx} distribution is compared for the isothermal calculation at T_0 and the non-isothermal calculation. The top half of the figure shows the contour lines of the isothermal calculation, while the bottom shows the non-isothermal result. Ten equally spaced contour lines between minimum and maximum values of the *isothermal* values are shown in *both* halves of the figure. Only in the regions with a significant temperature rise, are the stress levels altered.

Fig. 6 and 7 compare the isothermal with the non-isothermal dimensionless τ_{xx^*} component along the cylinder boundary as a function of the polar angle and along the symmetry line, respectively. The non-isothermal maximum stress level is about 20 % lower than in the isothermal case.

6. Conclusions

An algorithm to compute non-isothermal viscoelastic flows is presented. The method is based on the use of the Discontinuous Galerkin method to discretize the constitutive equation. So to satisfy the compatibility requirements of the velocity-stress discretization and to obtain a Total Variation Diminishing advection scheme, a constant stress element is used on triangles. Use is made of the Galerkin Least Squares

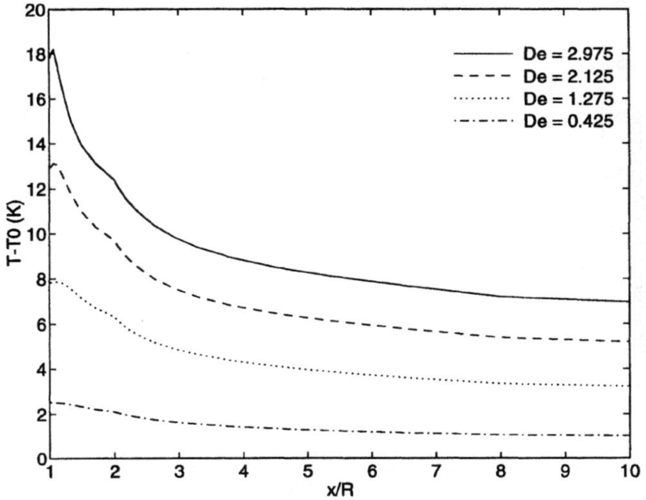

Fig. 4. The temperature rise $T - T_0$ along the symmetry line in the wake of the cylinder at four different values of the Deborah number as indicated in the figure.

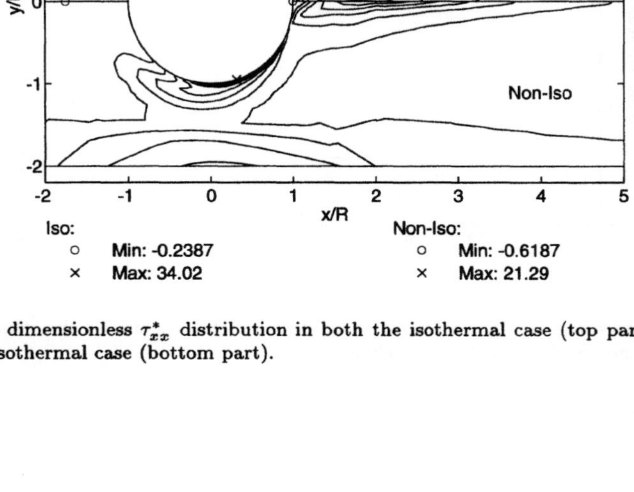

Fig. 5. The dimensionless τ_{xx}^* distribution in both the isothermal case (top part of the figure) and the non-isothermal case (bottom part).

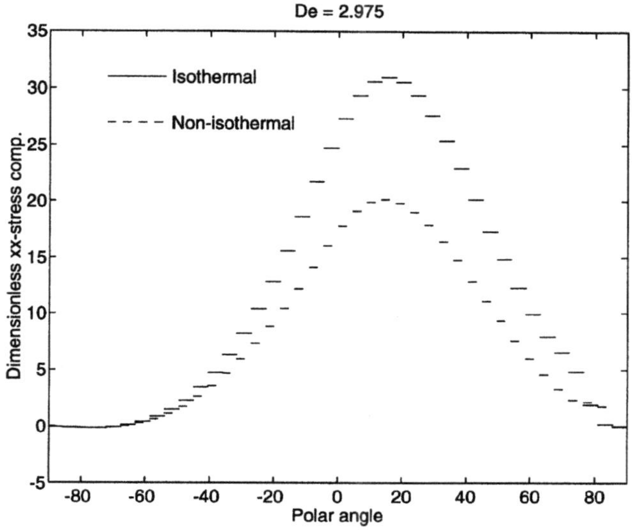

Fig. 6. The dimensionless τ^*_{xx} distribution along the cylinder surface. Both the isothermal and non-isothermal result are shown.

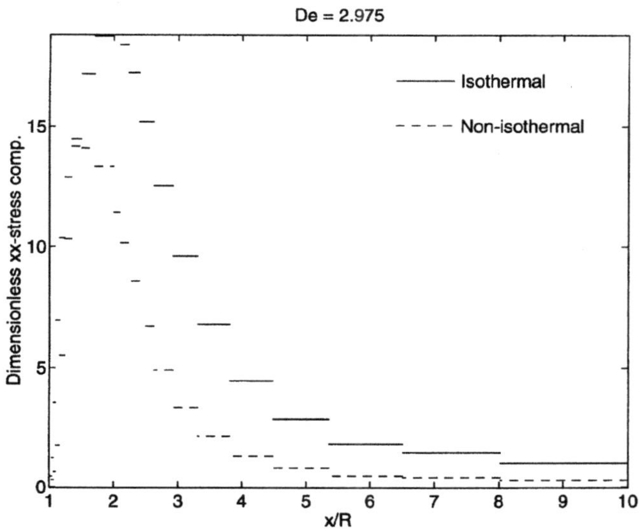

Fig. 7. The dimensionless τ^*_{xx} component along the symmetry line behind the cylinder. Both the isothermal and non-isothermal result are shown.

methodology to obtain stable, oscillation free pressure distributions.

The analysis of the non-isothermal viscoelastic flow of polystyrene melt around a cylinder modelled with the Phan-Thien Tanner model, shows that a significant temperature rise may be expected at moderate values of the Deborah number under the operating conditions investigated here. This temperature rise has a significant effect on the stress distribution in the wake of the cylinder. This clearly demonstrates that non-isothermal effects must be taken into account when evaluating experimental results.

References

1. F.P.T. Baaijens, H.P.W. Baaijens, G.W.M. Peters, and H.E.H. Meijer. An experimental and numerical investigation of a viscoelastic flow around a cylinder. *J. Rheology*, 38:351–376, 1994.
2. L. F. A. Douven. *Towards the Computation of Properties of Injection Moulded products: flow and thermally induced stresses in amorphous thermoplastics.* PhD thesis, Eindhoven University of Technology, 1991.
3. F.P.T. Baaijens. Application of low-order discontinuous galerkin methods to the analysis of viscoelastic flows. *J. of Non-Newtonian Fluid Mech.*, 52:37–57, 1994.
4. M. Fortin and A. Fortin. A new approach for the FEM simulation of viscoelastic flows. *J. of Non-Newtonian Fluid Mech.*, 32:295–310, 1989.
5. A. Fortin, A. Zine, and J-F. Agassant. Computing viscoelastic fluid flow problems at low cost. *J. of Non-Newtonian Fluid Mech.*, 42:1–8, 1992.
6. L. Franca and R. Stenberg. Error analysis of some Galerkin-Least-Squares methods for the elasticity equations. *SIAM Journal of Numerical Analysis*, 28:1680–1697, 1991.
7. A.N. Brooks and T.J.R. Hughes. Streamline upwind / petrov galerkin formulations for convection dominated flows with particular emphasis on the incompressible Navier-Stokes equations. *Comp. Meth. in Appl. Mech. and Eng.*, 32:199–259, 1982.
8. G.W.M. Peters. Thermorheological modelling of viscoelastic materials. In *Proc. of the IUTAM meeting on Numerical Analysis of Non-Isothermal Viscoelastic Flows, Rolduc, The Netherlands*, 1994.
9. X.L. Luo and R.I. Tanner. A pseudo-time integral method for non-isothermal viscoelastic flows and its application to extrusion problems. *Rheol. Acta*, 26:499–507, 1987.
10. O. Pironneau. *Finite element method for fluids.* John Wiley & Sons, New York, 1989.
11. J.M. Marchal and M.J. Crochet. A new mixed finite element for calculating viscoelastic flow. *J. of Non-Newtonian Fluid Mech.*, 26:77–114, 1987.
12. J. Baranger and D. Sandri. A formulation of Stokes' problem and the linear elasticity equations suggested by the Oldroyd model for viscoelastic flow. *Mathematical Modelling and Numerical Analysis*, 26:331–345, 1992.
13. R.C. King, M.R. Apelian, R.C. Armstrong, and R.A. Brown. Numerically stable finite element techniques for viscoelastic calculations in smooth and singular domains. *J. of Non-Newtonian Fluid Mech.*, 29:147–216, 1988.
14. D. Rajagopalan, R.C. Armstrong, and R.A. Brown. Calculation of steady viscoelastic flow using a multimode Maxwell model: application of the explicitly elliptic momentum equation (EEME) formulation. *J. of Non-Newtonian Fluid Mech.*, 36:135–157, 1990.
15. D. Rajagopalan, R.C. Armstrong, and R.A. Brown. Finite element methods for calculation of steady viscoelastic flow using constitutive equations with a Newtonian viscosity. *J. of Non-Newtonian Fluid Mech.*, 26:159–192, 1990.
16. P.J. Coates, R.C. Armstrong, and R.A. Brown. Calculation of steady-state viscoelastic flow through axisymmetric contractions with the EEME formulation. *J. of Non-Newtonian Fluid Mech.*, 42:141–188, 1992.
17. X.L. Luo and R.I. Tanner. A decoupled finite element streamline-upwind scheme for viscoelastic flow problems. *J. of Non-Newtonian Fluid Mech.*, 31:143–162, 1989.
18. R.C. Armstrong M.R.Apelian and R.A. Brown. Impact of the constitutive equation and singularity on the calculation of stick-slip flow: the modified upper-convected Maxwell model (MUCM). *J. of Non-Newtonian Fluid Mech.*, 27:299–321, 1988.

19. F.P.T. Baaijens. Numerical analysis of unsteady viscoelastic flow. *Comp. Meth. Appl. Mech. Engrg.*, 94:285–299, 1992.
20. R.I. Tanner and H. Jin. A study of some numerical viscoelastic schemes. *J. of Non-Newtonian Fluid Mech.*, 42:171–196, 1992.
21. F.P.T. Baaijens. Numerical experiments with a Discontinuous Galerkin method including monotonicity enforcement on the stick-slip problem. *J. of Non-Newtonian Fluid Mech.*, 51:141–159, 1994.
22. B. van Leer. Towards the ultimate conservative difference scheme. IV a new approach to numerical convection. *J. Comp. Physics*, 23:276–299, 1977.
23. Hughes T.J.R., L.P. Franca, and M. Balestra. A new finite element formulation for computational fluid dynamics: V. circumventing the Babuska-Brezzi condition: A stable Petrov-Galerkin formulation of the Stokes problem accomodating equal-order interpolations. *Comp. Meth. in Appl. Mech. and Eng.*, 59:85–99, 1986.
24. L. Franca and T.J.R. Hughes. Two classes of mixed finite element methods. *Comp. Meth. in Appl. Mech. and Eng.*, 69:89–129, 1988.
25. T.E. Tezduyar, S. Mittal, S.E. Ray, and R. Shih. Incompressible flow computations with stabilized bilinear and linear equal-order-interpolation velocity-pressure elements. *Comput. Methods Appl. Mech. Engrg.*, 95:221–242, 1992.
26. M.A. Behr, L.P. Franca, and T.E.Tezduyar. Stabilized finite element methods for the velocity-pressure-stress formulation of incompressible flows. *Comput. Methods Appl. Mech. Engrg.*, 104:31–48, 1993.

INDEX

INDEX

A

Andrade shift factor, 39
anisotropic heat conduction, 11, 37, 39, 54
Arrhenius law, 78, 104
axisymmetrical 4 to 1 contraction, 37

B

Bénard problem, 73, 76
Baaijens F.P.T., 179
Bazin V., 73
biaxial elongation, 14
Biot
  condition, 101, 115, 131, 159, 164
  number, 159
  parameter, 137
  problem, 131
Boussinesq approximation, 74
Brownian motion, 17
Brule B.H.A.A. van den, 11

C

capillary
  flow, 102
  viscometry, 102, 115
Carreau–Yasuda model, 135
Choleski method, 138
Chorin's scheme, 165
Clausius–Duhem inequality, 21, 27, 29
classical rubber elasticity, 25
constitutive equation, 2, 21, 22, 31, 38
contraction axisymmetrical 4 to 1, 37
corotational Maxwell model, 25
correction of the internal deformation tensors, 44
Couette flows, 80
Cross model, 126
Crouzeix–Raviart element, 42
cup-shaped mould, 149
Curtiss–Bird theory, 2

D

Davies A.R., 159
differential constitutive equations, 73
Ding D., 133
Dirichlet condition, 164
dissipation, 37, 40
Doi–Edwards
  differential approximation model, 25
  model, 21, 32
dumbbell, 2, 23

E

eccentric cylinder, 159
elastic deformations, 85
elongational viscosity, 4
energy
  elasticity, 21
  equation, 30
entropy
  elasticity, 21
  inequality, 27
experimental results, 85, 97

F

feed back control, 57
FENE models, 23
Fillon M., 85
finite
  difference discretization, 89, 111
  elements, 181, 37
flow
  around a cylinder, 179
  induced stresses, 154
  regimes, 65
Frêne J., 85
friction torque, 85
front tracking, 133

G

Galerkin method, 134, 138, 179
  discontinuous, 179, 180
Gauss–Seidel method, 89

Giesekus model, 2, 3, 6, 18, 26, 38, 41, 43, 47
Guillopé C., 73
Gunter S., 101

H

heat
  conduction tensor, 39
  flux, 39
Helmholz free energy, 30
high speed polymer processing, 11
Hopf bifurcation 57, 57, 58, 69
Hulsen M.A., 37
hydrodynamic instability, 73

I

injection moulding, 11, 133
inverse problem, 101

J

Johnson–Segalman
  –Oldroyd model, 57–59
  model, 25
journal bearing lubrication, 159

K

kinetic theory, 11

L

Langevin equation, 3, 17
Larson
  irreversible partial extending model, 25
  partial extended model, 22, 25
  simplified partial extending model, 25
least squares method, 127
Leonov model, 26
Li X.K., 159
Lim–Schowalter observations, 57–59, 61
lower convected Maxwell model, 25
lubrication systems, 11
Lyaponov-type identity, 61

M

macroscopic time reversal, 41
Malkus D.S., 57
master curve, 103, 122
material
  parameters, 104
  properties, 48, 93, 137, 167, 184
Maxwell–Oldroyd model, 157
microstructure, 32
  cup-shaped, 149
  T-shape, 148
multi-mode upper convected Maxwell model, 39

N

network
  model, 18
  theory, 11
Neumann boundary condition, 108, 144
Nohel J.A., 57
nonisothermal, 133, 179
  effects, 154
  dumbbell, 16
  elongational flow, 1, 3
  shear flow, 1, 5
  viscoelastic flow, 159
  viscosity, 154
nonmonotone, 57
non-Newtonian, 133
  fluid, 57

O

Oldroyd model, 83
oscillations, 57

P

particle tracking, 133, 144
periodic
  attachment and reattachment, 57
  orbits , 57, 69
Peterlin approximation, 2
Peters G.W.M., 21
Phan-Thien–Tanner model, 26, 44, 157, 179–181, 189
Phillips T.N., 101

piston driven flow
  with a contraction, 145
  shear flow, 57, 59, 60, 63
Plohr B.J., 57
polymer melts, 1, 179, 133
pseudo time, 1
pseudospectral method, 159, 165
purely elastic instability, 73

Q

quadratic systems, 57

R

relaxation
  phenomena, 16
  time, 163
reversed flow, 89
Robbins condition, 108, 159, 164
Rouse model, 23

S

Saut J.-C., 73
shear
  flow, 57
  flow correction, 44
  viscosity, 125
shift factors, 37
shifted power-law model, 135
Smoluchowsky equation, 23
source propagation, 133, 142
spurt, 57, 64
stabilization, 179
stepwise limitations, 44
stick-slip phenomenon, 57
streamline
  integration method, 37, 41
  upwind Petrov–Galerkin method, 45, 180
stress
  relaxation , 24
  flow induced, 154
  -optical rule, 20
  -thermal relation, 18

T

T-shape mould, 148
thermal conductivity tensor, 18
thermodynamics, 27
thermo-elasto-hydrodynamics, 85
thermo-rheological simplicity, 1
tilting-pad journal bearing, 85
time-splitting scheme, 110
time-temperature superposition, 1, 2
total variation diminishing advection scheme, 189
Townsend P., 133

U

uniaxial elongation, 14
upper convected Maxwell model, 25

V

Van der Waals interactions, 12
viscoelastic flow, 179
viscous
  dissipation, 11, 101, 119, 131
  heating, 73, 78

W

Wapperom P., 37
Warner force law, 23
Webster M.F., 133
Weissenberg–Rabinowitsch correction, 102
White–Metzner model, 26, 159–162
Wiest J.M., 1

Mechanics

FLUID MECHANICS AND ITS APPLICATIONS

Series Editor: R. Moreau

Aims and Scope of the Series

The purpose of this series is to focus on subjects in which fluid mechanics plays a fundamental role. As well as the more traditional applications of aeronautics, hydraulics, heat and mass transfer etc., books will be published dealing with topics which are currently in a state of rapid development, such as turbulence, suspensions and multiphase fluids, super and hypersonic flows and numerical modelling techniques. It is a widely held view that it is the interdisciplinary subjects that will receive intense scientific attention, bringing them to the forefront of technological advancement. Fluids have the ability to transport matter and its properties as well as transmit force, therefore fluid mechanics is a subject that is particularly open to cross fertilisation with other sciences and disciplines of engineering. The subject of fluid mechanics will be highly relevant in domains such as chemical, metallurgical, biological and ecological engineering. This series is particularly open to such new multidisciplinary domains.

1. M. Lesieur: *Turbulence in Fluids.* 2nd rev. ed., 1990 ISBN 0-7923-0645-7
2. O. Métais and M. Lesieur (eds.): *Turbulence and Coherent Structures.* 1991 ISBN 0-7923-0646-5
3. R. Moreau: *Magnetohydrodynamics.* 1990 ISBN 0-7923-0937-5
4. E. Coustols (ed.): *Turbulence Control by Passive Means.* 1990 ISBN 0-7923-1020-9
5. A.A. Borissov (ed.): *Dynamic Structure of Detonation in Gaseous and Dispersed Media.* 1991 ISBN 0-7923-1340-2
6. K.-S. Choi (ed.): *Recent Developments in Turbulence Management.* 1991 ISBN 0-7923-1477-8
7. E.P. Evans and B. Coulbeck (eds.): *Pipeline Systems.* 1992 ISBN 0-7923-1668-1
8. B. Nau (ed.): *Fluid Sealing.* 1992 ISBN 0-7923-1669-X
9. T.K.S. Murthy (ed.): *Computational Methods in Hypersonic Aerodynamics.* 1992 ISBN 0-7923-1673-8
10. R. King (ed.): *Fluid Mechanics of Mixing.* Modelling, Operations and Experimental Techniques. 1992 ISBN 0-7923-1720-3
11. Z. Han and X. Yin: *Shock Dynamics.* 1993 ISBN 0-7923-1746-7
12. L. Svarovsky and M.T. Thew (eds.): *Hydroclones.* Analysis and Applications. 1992 ISBN 0-7923-1876-5
13. A. Lichtarowicz (ed.): *Jet Cutting Technology.* 1992 ISBN 0-7923-1979-6
14. F.T.M. Nieuwstadt (ed.): *Flow Visualization and Image Analysis.* 1993 ISBN 0-7923-1994-X
15. A.J. Saul (ed.): *Floods and Flood Management.* 1992 ISBN 0-7923-2078-6
16. D.E. Ashpis, T.B. Gatski and R. Hirsh (eds.): *Instabilities and Turbulence in Engineering Flows.* 1993 ISBN 0-7923-2161-8
17. R.S. Azad: *The Atmospheric Boundary Layer for Engineers.* 1993 ISBN 0-7923-2187-1
18. F.T.M. Nieuwstadt (ed.): *Advances in Turbulence IV.* 1993 ISBN 0-7923-2282-7
19. K.K. Prasad (ed.): *Further Developments in Turbulence Management.* 1993 ISBN 0-7923-2291-6
20. Y.A. Tatarchenko: *Shaped Crystal Growth.* 1993 ISBN 0-7923-2419-6

Kluwer Academic Publishers – Dordrecht / Boston / London

Mechanics

FLUID MECHANICS AND ITS APPLICATIONS

Series Editor: R. Moreau

21. J.P. Bonnet and M.N. Glauser (eds.): *Eddy Structure Identification in Free Turbulent Shear Flows.* 1993 ISBN 0-7923-2449-8
22. R.S. Srivastava: *Interaction of Shock Waves.* 1994 ISBN 0-7923-2920-1
23. J.R. Blake, J.M. Boulton-Stone and N.H. Thomas (eds.): *Bubble Dynamics and Interface Phenomena.* 1994 ISBN 0-7923-3008-0
24. R. Benzi (ed.): *Advances in Turbulence V.* 1994 ISBN 0-7923-3032-3
25. B.I. Rabinovich, V.G. Lebedev and A.I. Mytarev: *Vortex Processes and Solid Body Dynamics.* The Dynamic Problems of Spacecrafts and Magnetic Levitation Systems. 1994 ISBN 0-7923-3092-7
26. P.R. Voke, L. Kleiser and J.-P. Chollet (eds.): *Direct and Large-Eddy Simulation I.* Selected papers from the First ERCOFTAC Workshop on Direct and Large-Eddy Simulation. 1994 ISBN 0-7923-3106-0
27. J.A. Sparenberg: *Hydrodynamic Propulsion and its Optimization.* Analytic Theory. 1995 ISBN 0-7923-3201-6
28. J.F. Dijksman and G.D.C. Kuiken (eds.): *IUTAM Symposium on Numerical Simulation of Non-Isothermal Flow of Viscoelastic Liquids.* Proceedings of an IUTAM Symposium held in Kerkrade, The Netherlands. 1995 ISBN 0-7923-3262-8

Kluwer Academic Publishers – Dordrecht / Boston / London

MIX
Papier aus verantwortungsvollen Quellen
Paper from responsible sources
FSC® C105338

If you have any concerns about our products,
you can contact us on
ProductSafety@springernature.com

In case Publisher is established outside the EU,
the EU authorized representative is:
Springer Nature Customer Service Center GmbH
Europaplatz 3, 69115 Heidelberg, Germany

Printed by Libri Plureos GmbH
in Hamburg, Germany